JOHN O'CONNOR

CANALS, BARGES AND PEOPLE

JOHN O'CONNOR

CANALS, BARGES AND PEOPLE

ILLUSTRATIONS BY
THE AUTHOR

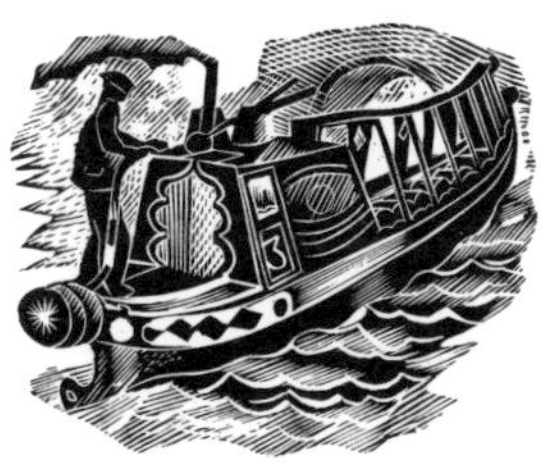

Unicorn Press Ltd

Unicorn Press Ltd
66 Charlotte Street
London
W1T 4QE
www.unicornpress.org

10 9 8 7 6 5 4 3 2 1

First published in 1950
Published by Unicorn Press Ltd 2014

ISBN 978-19100-65259

Cover designed by Felicity Price-Smith
Printed in India by Imprint Digital.

• CONTENTS •

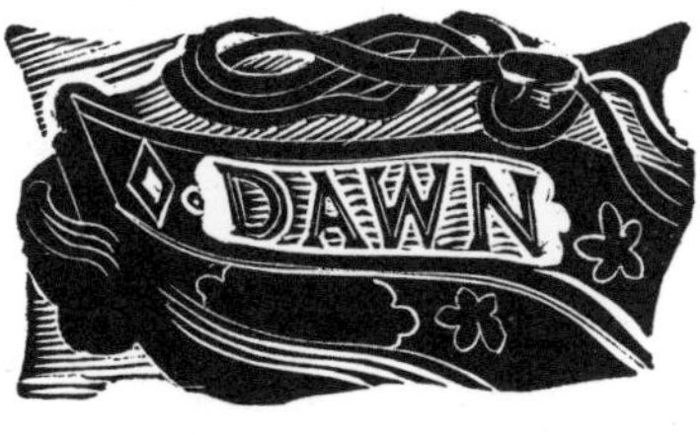

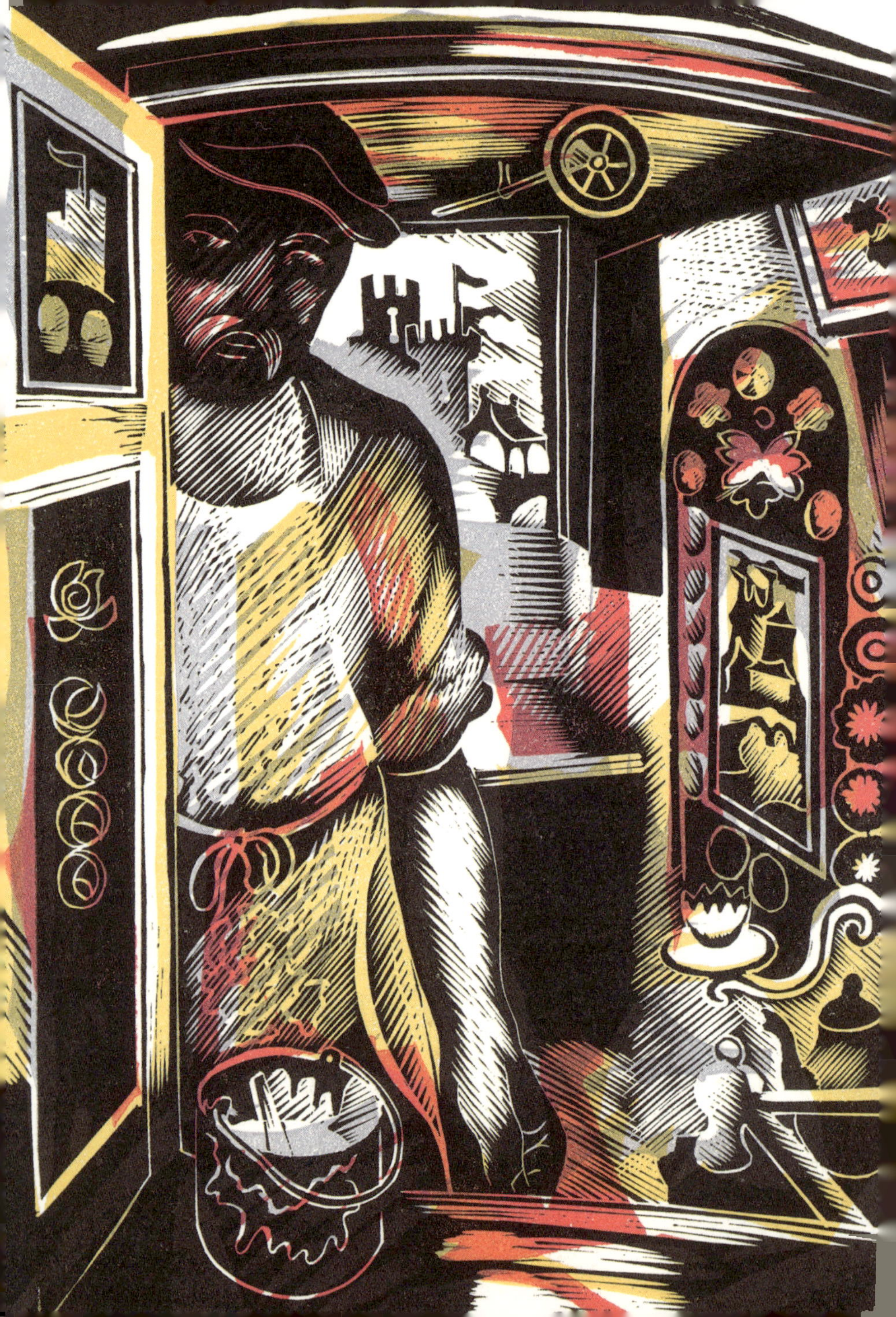

I

BEGINNINGS

It sometimes happens that works which are entirely artificial in their origin acquire an unintended beauty. Thus it is with the canals of Great Britain. There are some people who see beauty in a railway cutting or in the pattern of a railway junction, but the appeal is a limited one. The appeal of canals is wider, more akin to natural beauty – the charm of slow water, friendly bridges, locks, coloured boats and the small scurrying animals of the water bank.

By its very nature a canal waterway demands a prime consideration for levels, and the planner had to seek a level course whenever possible. This limitation endows a canal with a restfulness foreign to the high road, the by-pass and the searing railway embankment. There is a sense of leisurely repose in this simple, adequate way of getting from place to place by water.

To maintain a regular supply of water to the canals there are fixed points, provided in the initial engineering scheme, where water is brought by carrier stream from a lake or reservoir to the canal head. As a boy, I went often to the supply sluice of a Midland reservoir on the borders of Leicestershire and Lincolnshire, from which an artificial but very natural-seeming river flowed some seven or eight miles to feed the Grantham canal.

This stream roared madly through a strong iron grille at the opening of a tunnel about two hundred feet long, which was dark, damp and overgrown with ferns and flowers. Here the water bubbled with yellow foam, but over the deep sandy bed of the carrier it flowed at a steadier, saner pace. The thought of this stream brings back to me memories of summer days, of rough shirts, clean and papery-smelling in the hot sun, of baskets heavy with tomatoes, tea and big slices of new white bread on which whole jars of jam seemed to have been emptied; memories of laughter, sunshine and laziness, of an old horse, blind in one eye, with a straw hat and little tasselled covers made of yellow nets on his ears to keep off the flies.

Softly mossed stones, mould of twigs and leaf skeletons patterned the edge of an inland lake there, where moorhen, heron and crested grebe made their nests in the rushes. Along the side of the lake willows and chestnuts dipped deeply into the shadowy water, their roots laid bare by former floods. When there was a light wind over the water the tiny harbours and inlets between the trees were beaten by storms, dangerous to toy shipping.

We used to walk up to the higher ground behind the washed willow branches, avoiding beds of tall nettles. Here the water marigolds grew between silver-green stones of the overflow, where grass, laid low at the last high water, lay flat and dried in the sun.

Our path led us beneath a narrow, reed-filled bridge, dry except in the rainy season, to the cold moss-covered walls of a dry water course, between which we trod carefully from stone to stone.

Here, in the cavities of the largest boulders, slippery with the flowers of lime and elm from overhead, were pools of water two or three feet deep, left black from the last rains. Hartstongue ferns, brambles and low bracken, with sweet briar, catmint, bitter garlic and tall splendid foxgloves, grew on the bank above the stone side of the artificial gully. Daylight was almost hidden, and a strange, slightly alarming, cool greyness filled the world. From the tunnel a deep roaring rose to the trees, above the noise of flies.

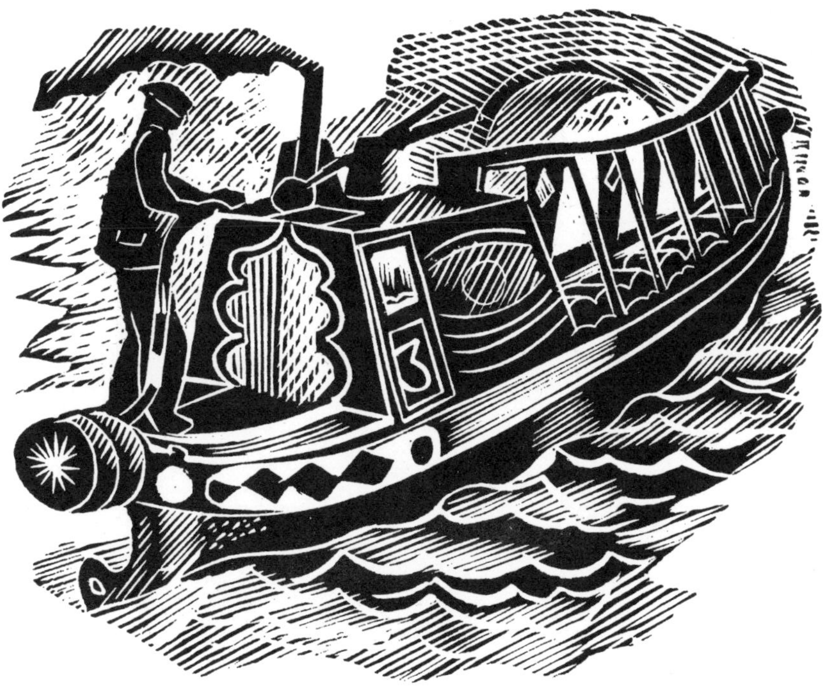

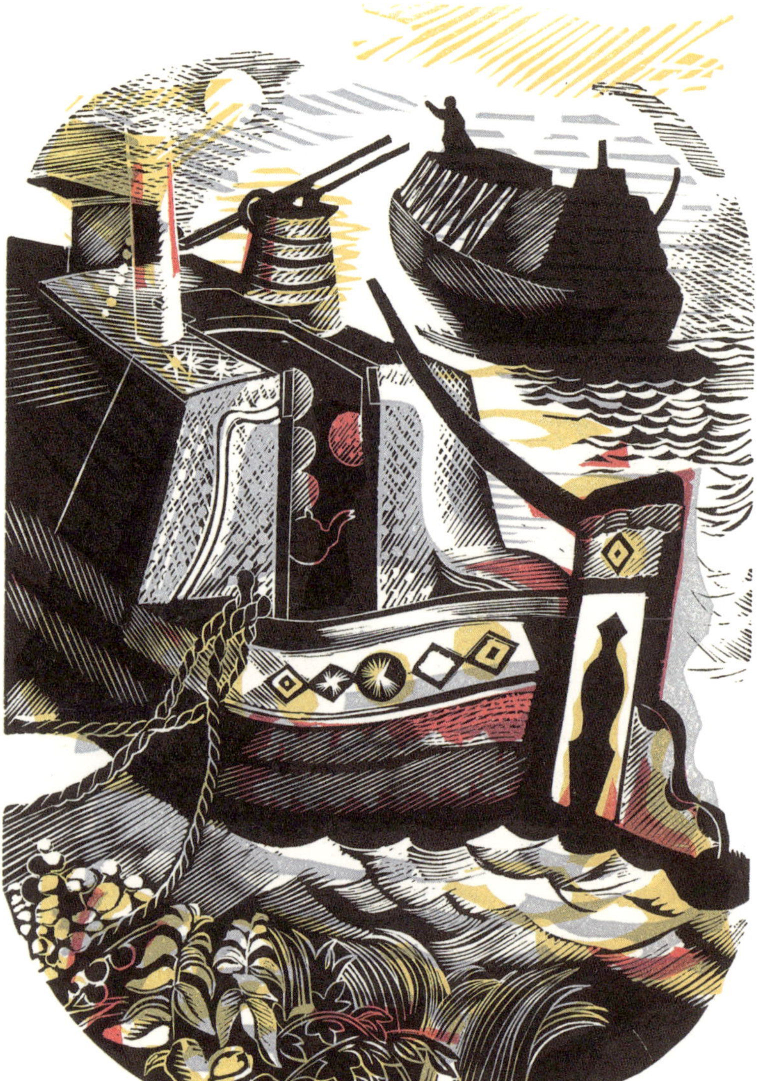

As we followed the waterway, the roar increased at each turn of the dripping path until we reached the first of a series of short culverted tunnels leading to the canal. The brick entrance was overgrown with brambles and elder bushes. Here it was that my uncle, when a youth of seventeen, had crawled on his hands and knees in two feet of water through the tunnel to rescue five prize ducks, due for show at a Nottingham fair the following day, which had waddled happily into the darkness, taken fright and remained. Later they appeared in a neighbouring field at the other end of the tunnel, three hundred yards from the home field, apparently none the worse for their travels; sufficiently unmoved, in fact, to carry away a prize at the show next day.

And so by tunnel or by path, one reaches the Grantham Canal, the first canal of many that I came to know. I could have had no better introduction. The Grantham Canal is short, but in the few miles of its length can be seen all the features most typical of an English canal.

Firstly, there are the stretches, hundreds of yards long, of rushes and weeds, through which no dredger has passed for many years, showing a few inches above water near the bank, and growing unseen beneath the surface in the deeper parts of the stream. Then there is the trodden patch at every field side where cattle come to drink and the lines of meadowsweet between the rushy water edge and the thick hedgerow at the side of the towpath.

The first brick bridge on the Grantham Canal sets the pattern for all the others, varied only by an occasional blue brick parapet or moulded stone quoin. The bricks are mostly narrow, being only two and a half inches deep, and brilliantly coloured by the yellow-green moss and efflorescent lichen which grows over them. Midland bricks, which remain hotter in colour than almost any others made in England, seem particularly to attract this glowing patchwork of mosses. The bridge itself has suffered settlements and additions until the edge of the parapet is as undulating as the furrows of the fields about it. It stands solid, warm and friendly, waiting for the gaily-painted monkey boats which no longer pass that way.

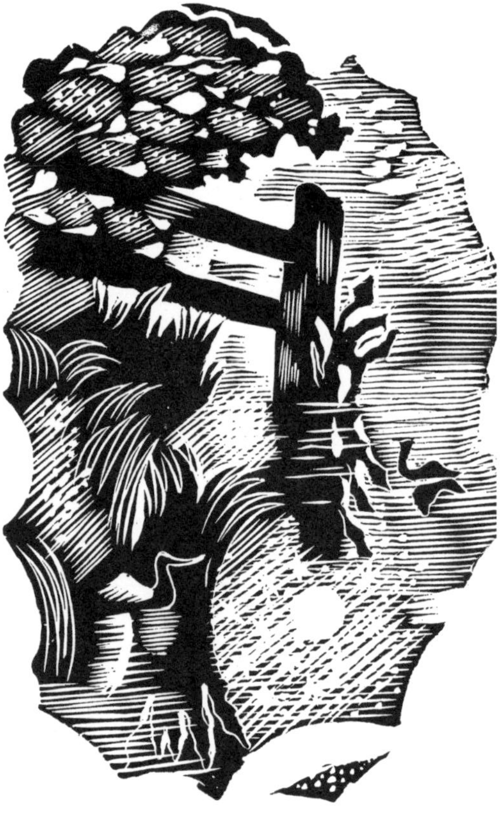

II

THE NEW MONKEY BOAT

Though the Grantham Canal is closed now, boats are still busy on other reaches, and new ones are always being built. So I went to Uxbridge, where there is a boat-building yard, and found a new monkey boat, which a few weeks before had been just a heap of sawn oak logs, nearly completed. Monkey boat, by the way, is the correct term for a canal dwelling-boat; a barge, or lighter, is a clumsier craft some seven or eight feet wider.

The first part of the boat's construction is a 60 to 70 foot keel made of sections of English oak from several whole trimmed trees with the bark left on. Around this keel is laid a flat base planking of elm from which the ribs are secured. There are ten permanent ribs in the prow and about seven each side in the hull which are only temporary. The keel board is raised five feet or so from the ground so that the elm base planking can be trimmed around the edges, as a cook trims pastry around a pie. The boat during construction rests on eight stands made from little turrets of timber laid evenly.

A single piece of oak forms the stem; to hold it in position it is scarfed to the keel by a strut of wood to the ground. The bow ribs are made in much the same way as those of a small fishing boat. Pieces of oak cut from curved branches give the prow its moulded shape, even before trimming.

The timber for this boat had been brought from Axminster. The oak was light in tone but well-seasoned, while the elm was green wood. The timbers are all shaped with an adze, which is a dangerous tool in unskilled hands; but a good craftsman working with an adze is a pleasure to watch. It is used with a swinging movement, the haft being held firmly against the body with the right hand and the left hand acting as a brake, for a slip would mean a bad cut on knee or leg.

The side walls are shaped on the hull base board to take a lining on the inside of thin oak boarding packed with oakum and to stop weeds adhering on the outside. Elements in the canal water help to harden the sap wood at the bottom of the boat, so that the inside lining of the hull keeps dry and snug for a great many years.

Canal boats, naturally, do not have to endure the tossing and vibration which a sea-going vessel must stand; but their extreme length and the immense loads they carry demand a very rigid frame, which the side walls, of one to two-foot planks, provide. About 40 feet of the boat is taken up by the open well of the hull, over which only a few tie-beams are used.

The most unusual piece of work in the boat is the large rounded stern, pierced by the tiller, into which a single massive piece of oak (roughly 21 inches wide and 21 inches deep) is inserted between

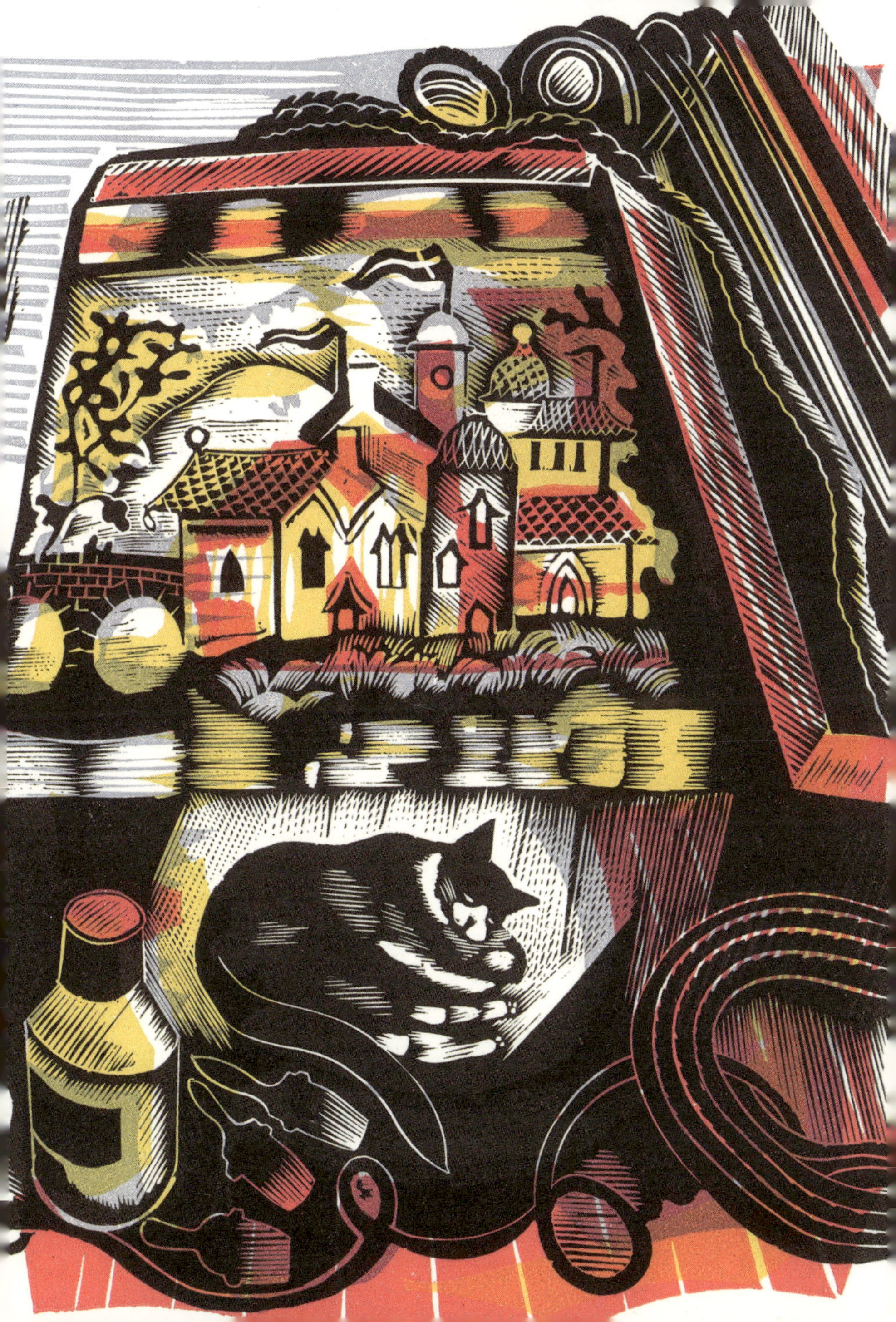

the planking to cover the tapering beam. The weight of this part is great in comparison with shorter craft, but in the motor-driven monkey boat great strength is needed around the stern post. The stern post itself, which is built immediately forrard of the large oak block of the stern, is a fine piece of oak about 5 feet tall and 15 inches wide, thickened below the water line in the shape of a plumber's joint for a boring to take the screw shaft.

No blue-prints were used by the workmen, although such prints do exist. They and the foreman, who has spent more than forty years in the employ of the Company, carry all the details of the construction in their minds. The structure of the wooden boat has remained unchanged for many generations and the work is built up piece by piece, on the principle, as the foreman said, that 'line and rule help many a fool'.

As soon as the prow ribs are completed and the breast hooks fitted between shelves and stern, the decks are constructed with huge carlines or shoulder pieces, shaped and fitted securely under the deck beams. This is done in exactly the same way; and the pieces of English oak used, though smaller, are exactly the same shape as in the construction of Nelson's Victory.

Before the dwelling cabin is fitted into position the whole of the interior is lined with oak, so that the stern portion inside looks like a huge hollowed tree. There are some fine shapes in this stern work, particularly a gradual curve from the parallel hull, running into an ogee terminated by the stern post.

The engine seating, cabin frame, roofs and hatches are now added, roughly painted with undercoating, so that a presentable-looking boat, only lacking tiller, central gangwalk, paint and trimmings, sits on its wooden turrets waiting to be launched.

The whole yard at Uxbridge took part in the launching, which was a complicated business, though the men assured me I should find it easy enough after ten years or so experience. First of all jacks were placed under the hull and raised to trestle height. Chains were threaded through pulleys, thrown over the boat, and fastened by a large hook. Then the jacks were raised a little and one by one blocks of wood were removed from the supporting turrets at the side nearest the water until the boat lay over at an angle of about 35 degrees. Next the jack- operated column holding up the centre of the roof was removed, so that the whole length of the boat was open to the water; the roof appeared perfectly secure without this additional support.

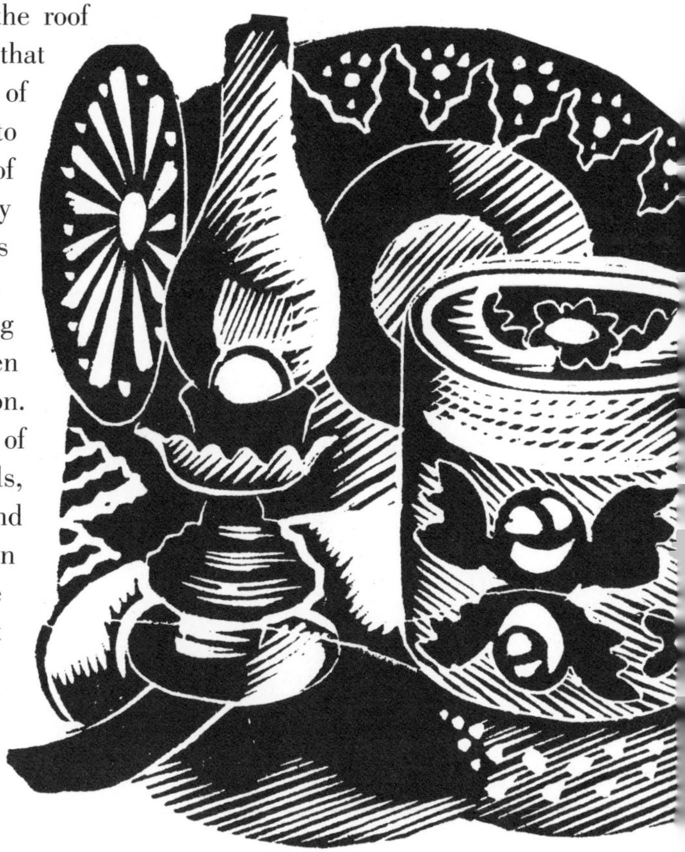

Two large sliding ramps were then placed into position. These consisted of two metals rails, tapered at each end and attached to an 18-inch square shaft of oak about 25 feet long. The jacks were now only about

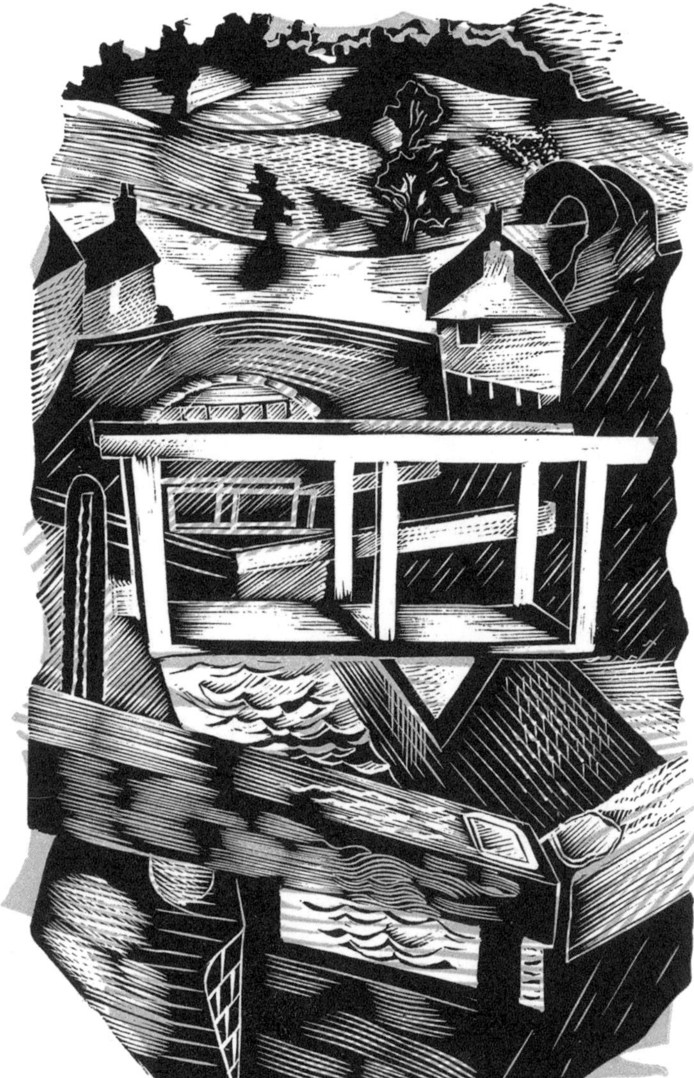

3 feet from the ground and the rail logs rested on transverse logs fitting beneath the hull. More blocks of wood were removed so that finally the boat and ramps rested on three jacks at an angle of about 30 degrees.

The chains were loosed on the pulley tackle and to cries of 'Steady!', 'Let her go', 'Steady now', 'Let her go!' the boat began to slide slowly down the greased rails towards the water. The rails were some 6 feet in the canal and there was no fear of the boat's capsizing, though she looked horribly likely to do so. But with a gentle splash she came easily to rest in the waterway where she would spend the next thirty or forty years. A wash of water flooded up as she settled, rustling the wood shavings on the floor of the building shed and stirring the reeds and rushes on the opposite bank.

This method of launching is hundreds of years old; it may have been used in that same locality by makers of flat-bottomed craft soon after the Norman Conquest. The same procedure is followed, in reverse, when a boat is taken out of the water for repairs. Chains are fastened round her hull and she is laid with her underwater edge resting against the iron rails, which are jacked at the land end and set at right angles to the water. The gang then hauls the boat up the rails, broadside to the canal. Wedges and jacks are put into position beneath her hull and she is raised to a workable height above the ground. Short wooden rollers are used if it is necessary to drag her farther inland.

Once launched, the new boat is ready for engine adjustments, decorating and additional fixtures. It is now that the painter comes into his own.

III

THE NEW MONKEY BOAT IS PAINTED

When I asked for the boat painter, the men said, 'Oh, you mean the new man.' This was, I discovered, because he had only been at the yard about ten years. They seemed to like teasing him about his floral cupboards and cheerful decorations. 'What do you suppose will happen to the flowers and patterns when Henry stops painting them?' asked the young man who helped him 'Because I'm not stopping on this job. They just won't get done, I reckon.' A boatman who was in the yard at the time said, 'If you'd ever lived in a monkey boat you'd think a bit different about them. I suppose if you give up we'll just have to do them ourselves.' I hope they will, for the canal would be a sadder place without the cheerful impudence of these painted boats.

I found the paint-shop next door to the little forge where the blacksmith deals with all the metal fittings. 'Paint-shop' seems an understatement – no words could describe the quantities of paint

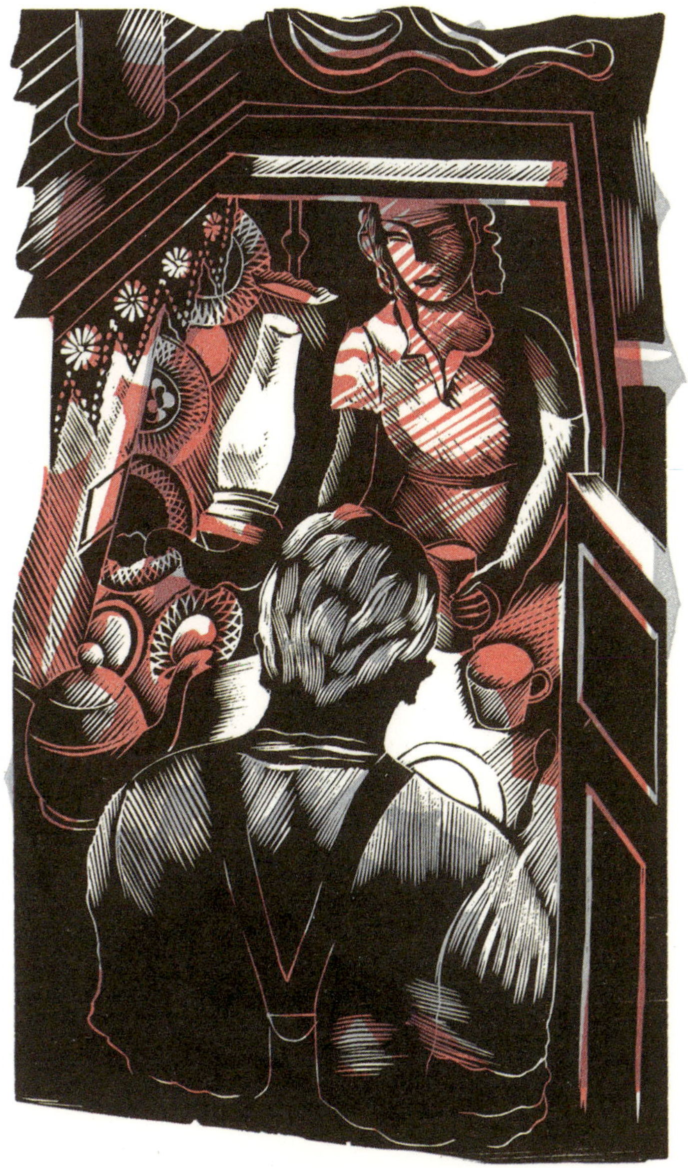

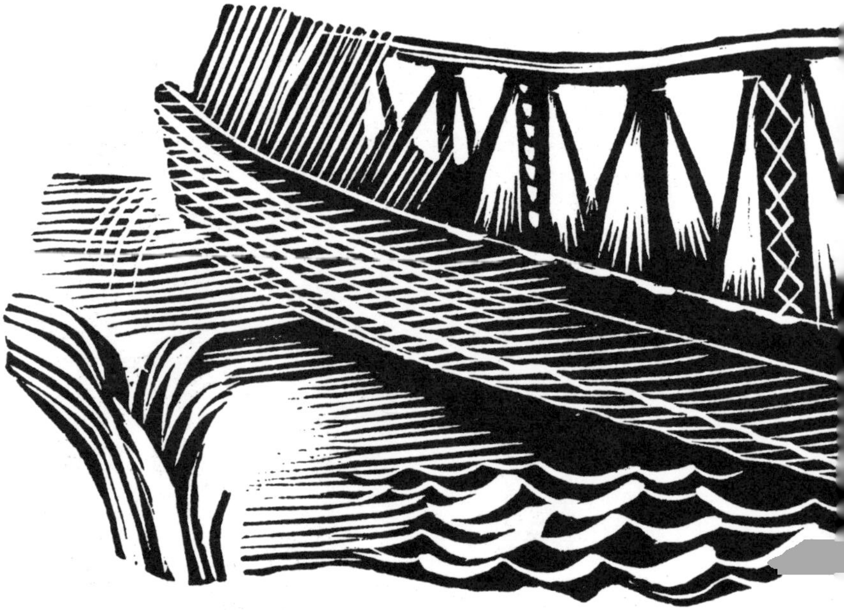

everywhere. Walls, floors, benches and shelves exuded paint. Colours have been mixed and tried for generations on the wooden walls, so that they bear a crust of scarlet, blue and yellow paint. Around the walls were tillers and tiller-bars, fore and aft hatch covers decorated with red hearts and diamonds, square blocks I did not recognize, but which I later found to be cordways for the lashing of tarpaulin, and innumerable seats chequered red, black and white. Along one side lay a boat's central gangplank patterned in four colours on green; in a corner was a fascinating set of little knife drawers; and among all his treasures stood the short stocky figure of Mr Penn the painter, sleeves rolled to the elbow, wide carpenter's fingers stirring treacly colour in a tin. His cloth cap, which almost covered his face, was a little darker than his grey

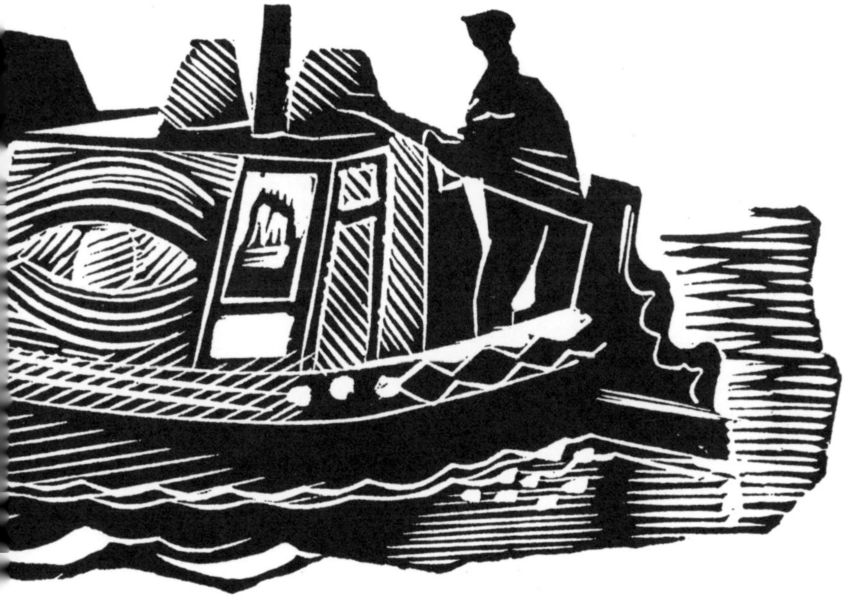

moustache and strangely dull among this array of colour.

We talked together about his painting. ‘Before 1914 we used the colours very little, and it was nearly all black and white,’ he said, ‘with gilded and mouldings.’ The designs and colours now used in this yard are quite its own. The bucket designs, side pictures and rose sprays painted at Braunston in the Midlands are irregular, seemingly derived from French rather than Eastern European designs, but the patterns at Uxbridge are formal and abstract. ‘They go in for decorative work up there,’ said Henry Penn. Nothing could be more decorative, to my mind, than his own work.

It is possible that the floral designs inside these boats are derived in part from the use of such designs for wallpapers and chintzes in the eighteenth century, towards the end of which boat dwelling was officially authorized. But there is a longer tradition behind them; for though the rather heavy roses displayed on boats hark back to early Victorian wallpaper and eighteenth-century brocades, the sprays of flowers also used have altered little since they were used by Elizabethan carvers and later by Georgian modellers in plasterer's jelly. Going farther afield, a panel from a door at Versailles or Fontaine-bleau might well be the elegant forebear of the floral cupboards of a canal boat.

In fact, for as long as man has been able to enjoy and remember the shapes of flowers the simple spray design has been used by almost all craftsmen. It is even possible that these decorations on panels and cabin doors are a faint echo in England of the floral festoons and vine-trailing jars of Graeco-Roman paintings.

The checks and diamonds of the central planks and helm seem to come from the cloth of Harlequin and, with him, go back to Byzantine origins, and are similar to Rumanian and Ukrainian decorations. The decorated household pottery and the great jugs and basins of the eighteenth and nineteenth centuries come into the canal scheme of things too; while the plates with china ribbons threaded into lattice-work, still proudly owned by the boat families, come in spirit from France by way of the potteries of Stourbridge and Worcester.

But to return to the paint-shop: it has a stove in it and in the winter months becomes the meeting-place of the men in the yard. They came in now, leaving their groups of tools by their working places in the boat, symbols of the many different jobs that produce the final result. 'A bit of fire keeps body and soul together these cold days,' said a tall lad. Around the stove the talk was not of boats, but of racing, football pools and gardens. 'A drop of your green paint there would come in handy, Henry, on the door of my shed,' said one, and 'You just spread them out like young cabbages,' explained a thickset carpenter by the door, until the foreman looked over the edge of his spectacles like a puzzled curate and said, 'Well, let's be going, lads.'

The men went back to work and I wandered around the boatyard, looking about me. The most unhappy sight was a sleeping group of old tillers standing against a shed wall, like a group of old village people huddled together in the thin warmth of the evening sun. Most of them were hopelessly broken, but all had been at one time very colourful above the waterline. Judging from the weeds and flowers that grew around these wooden relics, some even growing through the cracks between the jointings, they had leant there undisturbed for many years. Each bore a number, and somewhere in Great Britain there is, or was, a narrow boat bearing that number, chugging between villages or lying at some wharf-side. Boat crews are not fussy about such things, so that tillers and hatch covers and lashing blocks may change hands many times in the course of a few years. A woman called out to me from a boat one day 'That was our tiller, that one you were drawing last week. We might get it back one day. It's been painted, and now they're putting new irons on it.'

I went down to the canal-side. There were three boats in dock at the time, side by side, forming a complete contrast. The first was the most derelict and dilapidated object imaginable; she seemed hardly floatable. Her colour was a dull reddish-grey with blotches of dirty white where there had once been a pattern. The second was painted all over in a primary coat of dull pink, the decks and cabin looking like an unfinished doll's house waiting to have its

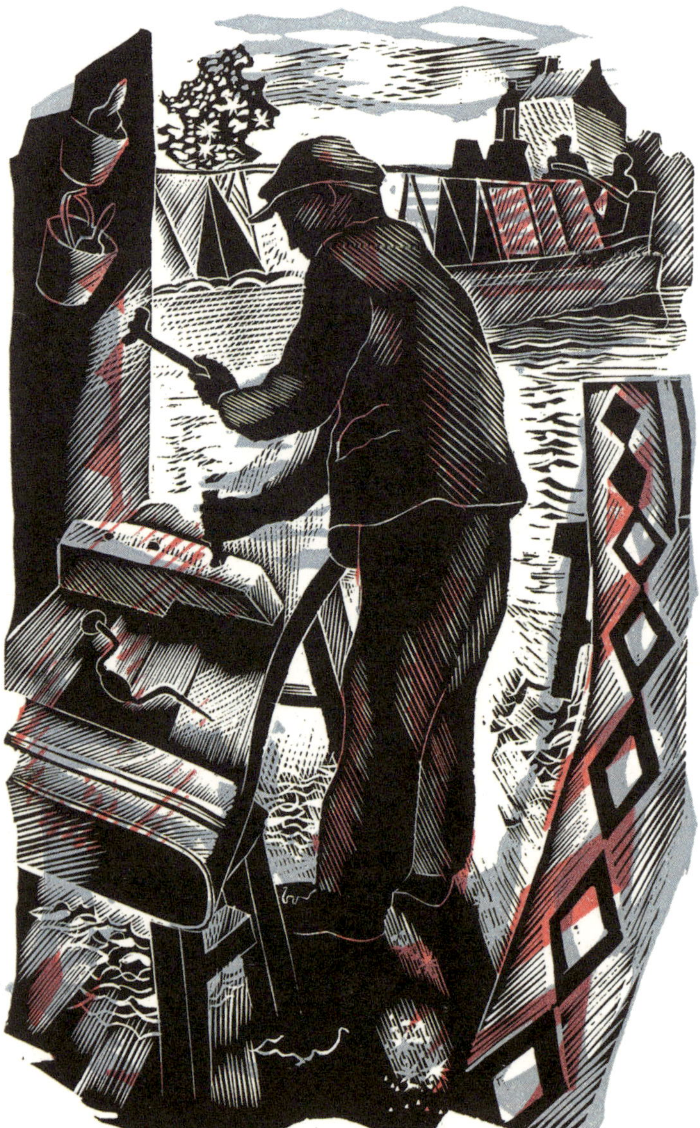

window curtains painted on. I was reminded of an old woman who had admired the curtains in a lock house one day. 'Why don't you have some of the same material in your boat?' I asked, forgetting that she had no windows. 'Ah, the stuff's all right,' she said, 'but it would look a bit silly, like, over the door.' However, later on she did fit a curtain over the door leading into the hull. But she still had a grumble. 'It looks grand,' she said, 'but it's a bit to keep clean.'

The third boat was the new one in all its glory. The sign writer was painting careful shadows behind the lettering, the iron work was black and shining, awaiting the fitting of the brasses, and the cords and plaited ropes on the helm were whitened and ready. The engine had been checked; I looked into its compartment and was surprised to find a fine floral decoration on the wall. It seemed strangely out of place, but very pleasing. A temporary tiller had been fitted which would very likely stay with the boat for some years; its own tiller stood in the yard leaning upon boxes and stakes. This dismantled tiller seemed abnormally large, and its decorated headband very small indeed. The main rudder was white with yellow edges, the pattern and metal bands being red, blue and black. Over the metal plates the carpenter fixes plaited bands of white cord. A new-painted canal boat carries on that traditional English love of a good turnout which still appears in the Lord Mayor's Show and in small-town holiday processions. The same spirit is to be seen in the draymen's show teams, the mane and tail of each shining horse gay with plaited ribbons, and in the colours of the Coronation or Jubilee souvenirs of the past eighty years.

The cabin of the new boat had neither people nor household things in it, and wore a dignified, impersonal air. There was no cheerful fire in the grate, and no brass piping, plates or tea-kettle for the firelight to flicker on. The table and pantry, not yet obscured by coats and crockery and food, could be plainly seen. It seemed a pity ever to cover up the little pantry shelves. In the half-circle above them was a pattern of roses, grouped around three single dahlias, the main design being the traditional arabesque. When shelves like these are painted the underpainting is first grained, then the flowers are added and lastly thin hair lines from the leaves, the whole usually against a ground of olive green. The outer mouldings of these coloured shapes are red. On a rougher scale, this cupboard-fitting is like the built-in shelves of a Regency house.

It occurred to me, looking at the door of such a cupboard, that the appearance of so many roses and balanced flowers in monkey boat cabins might possibly be a derivation to some extent from the work of William Billingsley, the nomad potter, and his pupils: the difference is, of course, that Mr Billingsley worked upon very expensive and highly valued materials, whilst the monkey boat decorations, though delightful, are of a cruder character, and not created with an eye on posterity. The similarity is most pronounced in the looser type of rose painting from Braunston. The more formal flowers of Uxbridge depend more upon a 'reflection' pattern, in which two or three roses are depicted back to back as if painted against a mirror, the repeat theme which was used so splendidly by the tapestry weavers of the seventeenth and eighteenth centuries.

The box decoration of Mr Penn is in a clearly defined process of painting, first the common background, marked with chalk for the position of the leaves and flowers. The second colour is a paler green for leaves, then the pink, white and yellow petals. Against these petals and leaves the black shadow strokes are painted. Dark veins and stems follow, and finally white highlights and yellow strokes.

Coastal barges of the type that are to be found at the quays of Rochester, the Thames and the near ports of the East coast, Maldon, Colchester and Ipswich, are not by nature of their work related to the canal craft, but there is one feature which is common to both. At the stern of the coastal barge great trouble is taken over the design and painting of the name of the vessel and its home port. Folded and flowing ribbons edged with gold and filled in with blue and red heraldic fields are most popular. Sometimes there are painted neat flower sprays beneath the ribbons and letters, large white daisies are common in these sprays, and the rose-like flower is very much akin to the rose of the canal boat cabin. I once saw two canal boats alongside a Thames barge, and the sturdy sea lines of the larger ship made the canal craft take on a strangely landlocked appearance. It was very clear that any similarity goes no farther than the decoration on the stern.

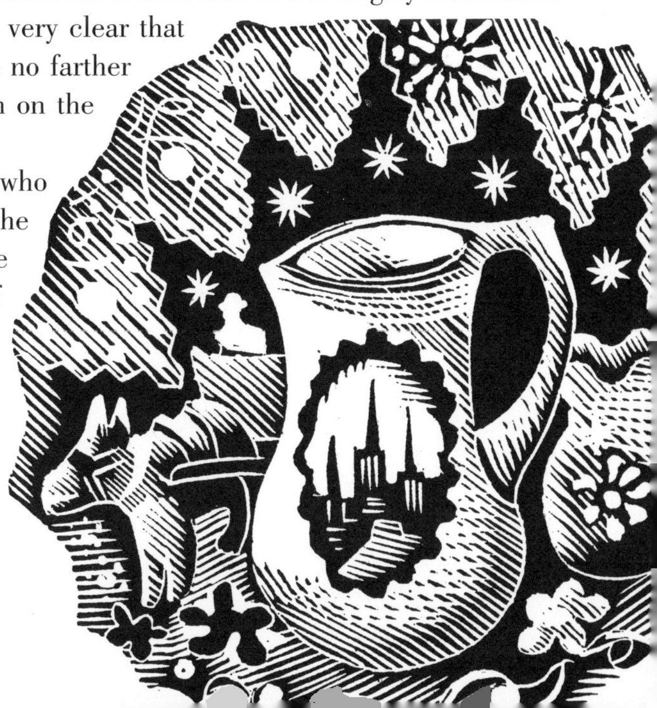

I met the man who was taking over the new boat. 'They've made a tidy job of her, I reckon,' he said. 'The missus will be right proud of the new paint.' I asked him if they were leaving the next day. 'Oh no, we never move off on a Friday,' he said. 'It's unlucky, you know, it just wouldn't do.'

Even a boat yard stops for tea, and immediately the clock on the tower chimed five newly-painted, half-finished and skeleton boats were all left deserted. They stood awkwardly in their setting of oak shavings, wedges, clamps and mallets. The only boat that did not look mournful was the newly-painted *Neston*. Gleaming with scores of flowers, red, blue and yellow diamonds along the green planks and castles with red-flagged towers against blue skies, she awaited the family who would occupy her.

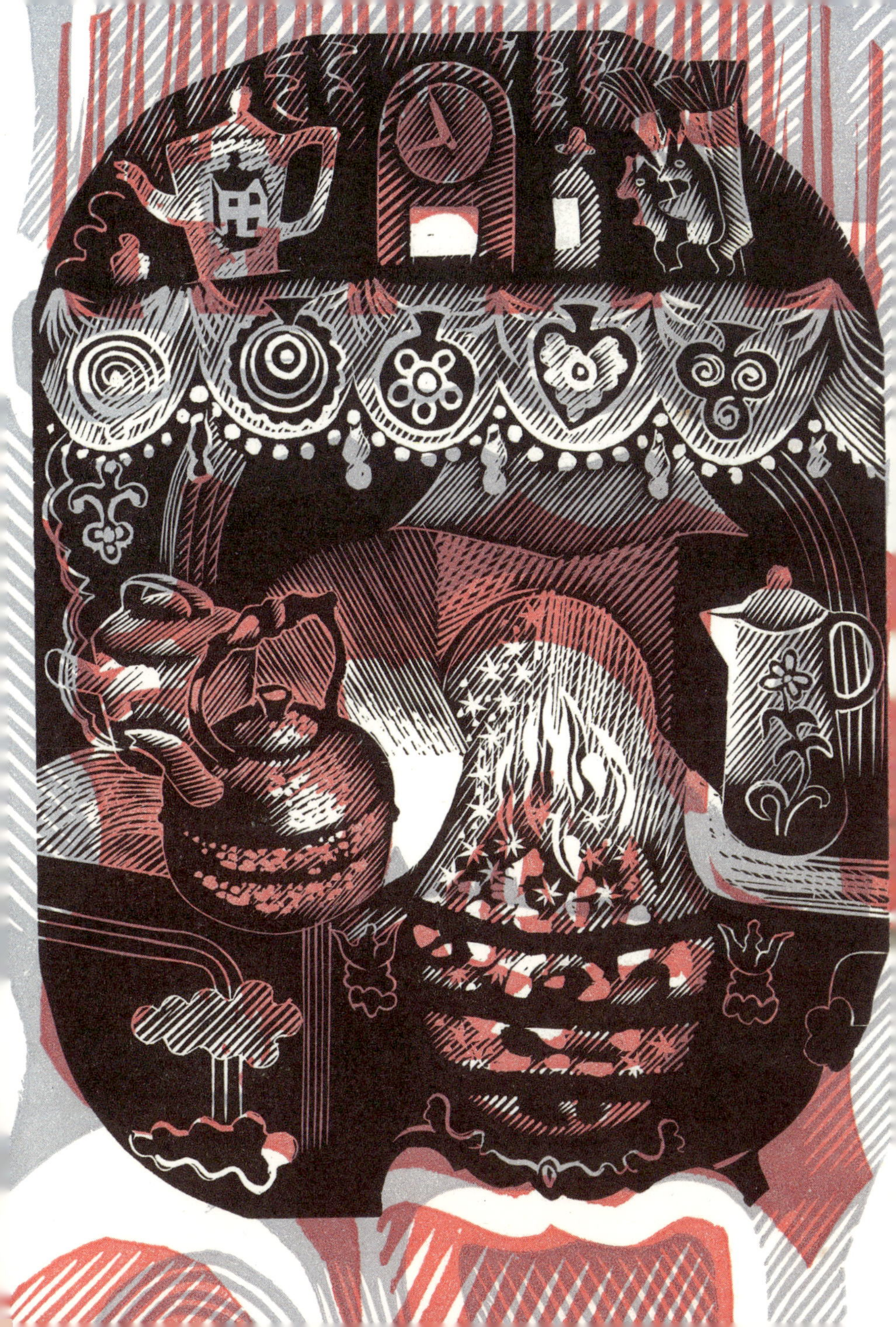

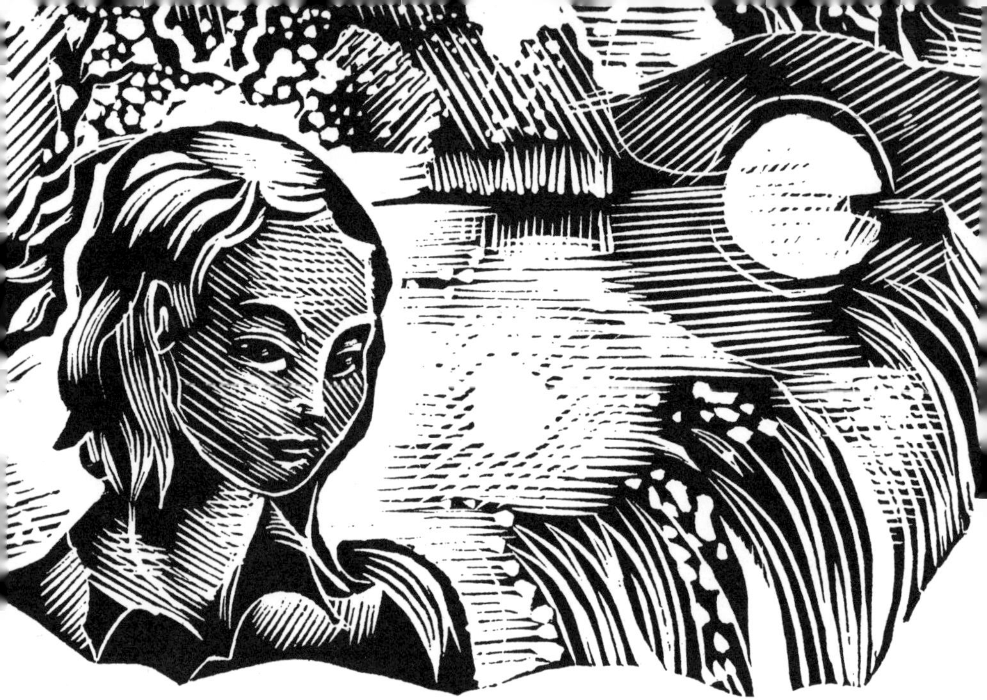

IV

THE FAMILY IN THE *NESTON*

The name does not really matter, and I will call them 'the family'. Joe is thirty-seven and was born in a riverside cottage in Hertfordshire; his wife was born in the cabin of a monkey boat thirty years ago. She is a short, jolly woman, becoming perhaps a trifle portly for the small hatch of the *Neston*.

He is lean, tanned and clean-shaven, with kind brown eyes and a slight stoop. 'I was for ever bumping my head on the cabin roof when I was a lad, and it makes you afraid to stand up,' he said. I first saw them aboard the *Neston*, wearing the characteristic black and green cord of canal folk, at ten o'clock one sunny morning at Brentford. She had been to buy the family rations ready for their journey. She stood on the towpath and called in a hoarse but pleasant voice, 'Joe,' across the motor-boat which lay between the dwelling-boat and the bank, protecting it from dogs and intruders.

Joe had been rolling his own cigarettes, which he preferred, leaving 'the brands' to the 'missus'; she liked to have one now and then as she stood at the tiller of the second boat while they were on the move. He came out from the cabin now and took the bag of rations from her. They were a happy and pleasant pair, with two children, Doreen, aged seven, and a thirteen-month-old baby. The sole disturbance on board the Neston that morning was created by the baby, who had two front teeth and was cutting some more; she sat, noisy and disgruntled, on the bunk of the cabin. 'She's a little terror this morning – it's her teeth. Don't you go teasing her again, Dorrie, she doesn't want any bothering. I'll have one of your packet cigarettes, Joe.'

Nevertheless, baby-minding appeals to canal mothers, for it relieves the occasional tedium of the journey. 'There's not much chance of dozing on the move, for there's a corner to turn or a bridge to be passed every few hundred yards,' said Mrs Joe, as we sat on the little deck outside the cabin door, where Doreen was mending a wooden steamer which had been towed on a string for hundreds of miles behind their boat.

Doreen attended the boat school from time to time at Brentford or Southall, usually the latter, where a big 15-foot-wide barge has been converted into a school. But her attendance card showed more entries at Northampton and Birmingham than in London. 'We

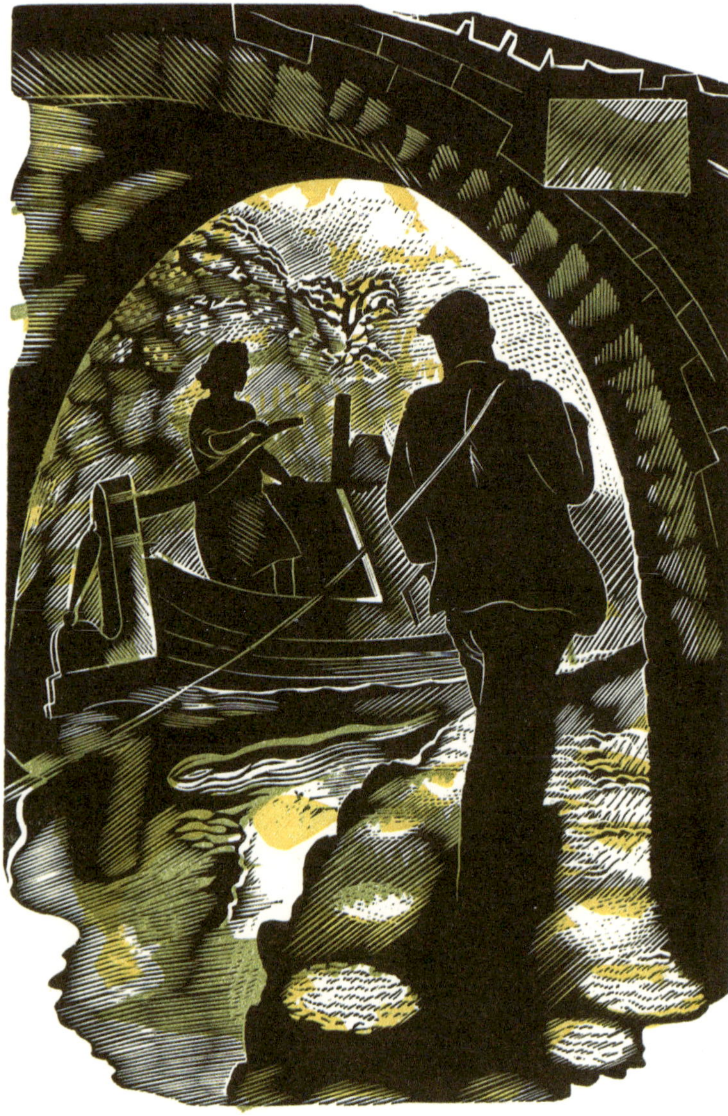

can't leave you behind here, Dorrie,' said her mother. 'If she comes with us, though,' interrupted Joe, 'she'll not be locking and boat-hauling. We can do without that, anyway, she's too young for it.'

There is not a dock or a leg of the main canal lines between Manchester and London that these people do not know. They are happiest when on the move and declare that they could never 'live on shore'. This seems a strange expression in a way, as much of their time is spent 'on shore' – in towpath walking, lock work and even shopping. They are a quiet people, as is to be expected from men and women who for years have moved slowly and peacefully along the waterways with time and opportunity to contemplate the late hours of sunshine and the early hours of day. They know and love the morning mists and the scent of hay over the wet fields. 'I shall be glad to be off,' said Joe. 'Give me the country any day. I can't seem to keep awake in this town air.'

Canal folk should never be called gipsies, for though, like them, they love the open skies and quiet hidden places, they have otherwise very little in common with the caravan people. The boatmen of Britain are generally placid hard-working men in their established trade, less nomadic than their European counterparts, living journey by journey, laying up a store of memories for the time when they will be too feeble to open the lock-gates themselves.

Joe loved the winter and early spring when the locks were frosty and long lines of spiders' webs glittered along the fences beside the water. 'There's more work to be done, with quick pay,' he said. 'And it's good of an evening to get back into the cabin, with the warm stove and the hatches closed and a pie waiting, after a long day on the towpath and hard lock duties on frozen sluice handles.'

The wife on a canal boat is just as house proud as her shore-dwelling counterpart in any suburb. She polishes the brasses on the outside of the boat rather in the same way as a Yorkshire housewife bricks her doorstep. Often there are unnecessary apologies for the pleasant natural disorder in the cabin. Mrs Joe would place her solid body squarely in the hatch, declaring 'the place wasn't fit to be seen'. The confined quarters means that the cleaning routine must be skilfully organized and extra precautions against disease must be taken. For instance, fresh milk is never carried on a boat; canal tea is always made with tinned.

The cabin was full of the treasured possessions of the family. Trinkets, medallions and cords, which formerly belonged to horse harness, were fastened to the chimney and to the stove. 'I polished them after we'd been settled a bit yesterday evening,' said Mrs Joe, 'and they're still as bright as when I left them. They're good brasses. They belonged to my mother, and she'd had them a bit.' They shone superbly on a white cord in the morning sunshine, bright enough to satisfy the most exacting parade sergeant-major.

'I grew up with the horses,' said Joe. 'They belonged to posting stables up and down the country, where they were changed or rested, and though we've got a good engine which doesn't need much attention, I'd rather be leading a horse. It was more matey like, and you were around the cabin more. You see, me and the missus spend the day now at the tillers of separate boats, so unless company orders close the locks early because of water shortage – they do sometimes between seven at night and six in the morning – we don't get a word in much. I decide the route and travelling time, so it's up to me really to arrange things as we like best.'

The *Neston* was going to Nottingham and then on to Peterborough before returning to London. She had headlamps for night travelling and electric light in the cabin from the same generators. Apart from lighting and engines, the appointments of a canal boat have altered little in the last 60 or 70 years. Joe said, 'There's not much risk of being stranded up the line nowadays, though, mind you, I can think of some pretty awkward places to get engine trouble, where it would mean a five- or ten-mile walk to the nearest telephone and a good bit farther to a village.'

Usually if a boat is laid up with engine trouble, she is towed by horse to the nearest depot or company dock where the load may have to be transferred to another boat if one is available. Urgent cargoes are not usually carried by canal boats, so that a delay of several days would not cause a serious inconvenience. The family might be taken to friends or relatives at a nearby lock house or village cottage. So many boat people have families scattered among the canal regions that to be isolated in some apparently remote spot seldom worries them, and usually affords great pleasure to the children who love an occasional holiday on land.

When they were in regular use horses were kept at special stables down the line, and fodder of a useful weight carried in the butty boat. Certain field and towpath pasture areas were allocated for grazing. There is a somewhat alarming feeling of inevitability about the slow but unbroken progress of a great barge horse moving along a narrow towpath. Subject often to a series of different masters, these heavy work horses were not always of the best of tempers. Anyone meeting a tall horse in a narrow cutting with a slippery towpath would be wise to make himself as small as possible. Horses are now used for short journeys only, wearing still the painted towing bar and diamond-painted nosebag of tin.

By five o'clock that evening the *Neston* would be several miles

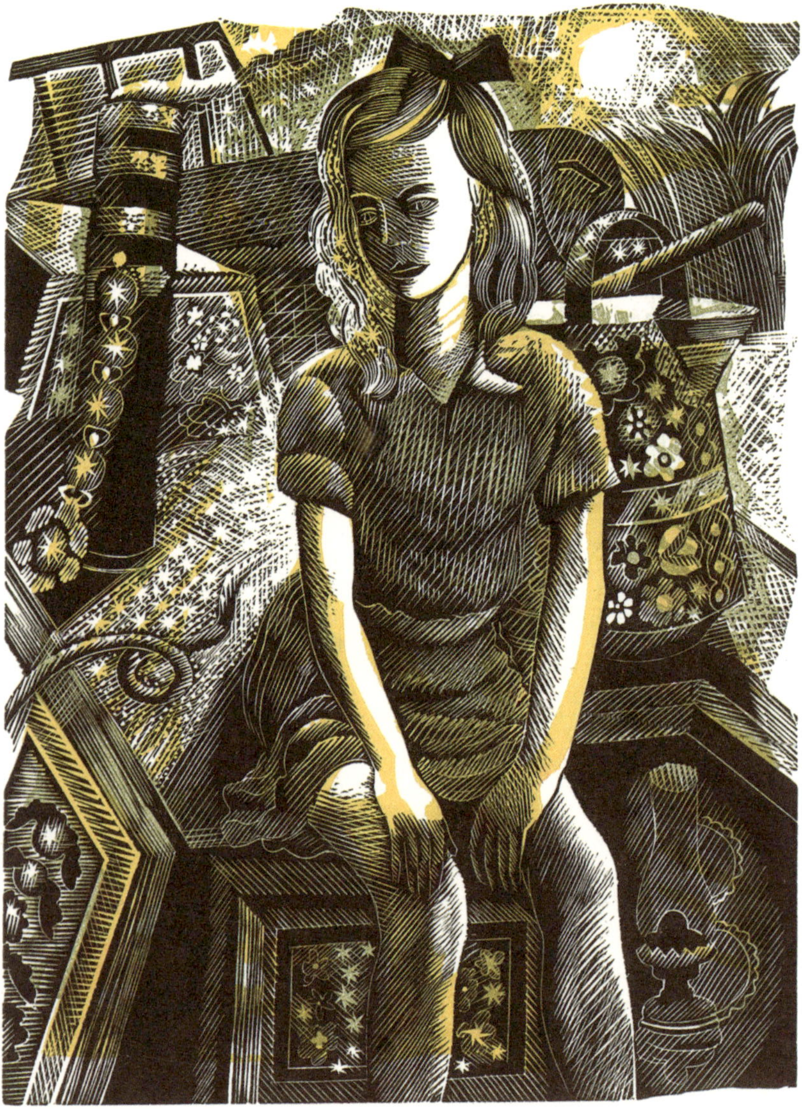

into the Home Counties. Dinner was finished and, with engine running steadily and a few last hurried remarks exchanged with other boats at the wharf, the *Neston* and her companion boat moved out of their position. With as little show as if they had been going for a summer afternoon excursion they headed for the first of the many locks that lay between Brentford and the wharves for which they were bound. 'I reckon we'll be away about a fortnight, or longer maybe,' called Joe as the motor boat rattled away.

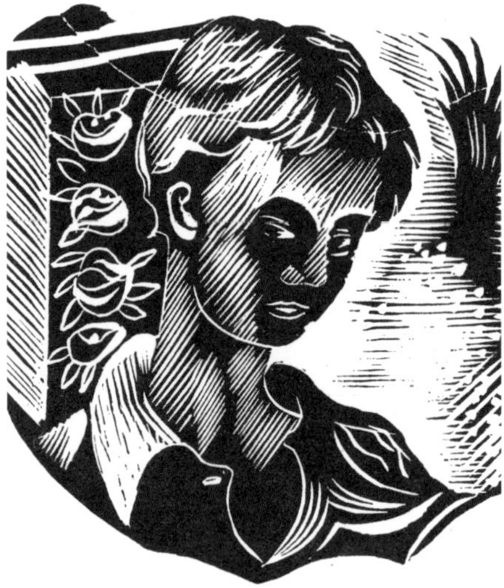

V

LOCKS AND MEALS

The landscape of the canals varies naturally as the watercourse passes from one county to another. In the Midlands it is clear that there has been considerable resistance from estate owners to the intrusion of the waterway, as with the railway, many years later. I recall a Leicestershire hamlet which in the year 1700 must have presented a very complete picture, but whose character today is quite changed by the passage of both canal and railway through the parish; it has now one of the most attractive locks and bridges in the county.

A tiny church with shallow steeple and ivy-grown tracery stands in its small graveyard. Across a lane is a Queen Anne house with a splendid walled garden. Two or three small houses and a farm previously completed the picture of Newton Harcourt. Now the canal runs along the border of the graveyard in a fine curve, leaving the walled garden undisturbed and a white lock house has joined the farm cottages of the hamlet. The hump-backed bridge connects the fields of Newton Harcourt with those of the neighbouring parish. Cattle crossing from the water meadows of Whistow Hall climb the steep approach of the brick bridge to their own farm on the other side. This is to me a typical prospect of an English canal. The curve of the watercourse reflects perfectly church, elms, and brick bridge in turn. On a hazy morning in late summer it seems that only an archaic Chinese painter could do justice to this scene; a thin mist lies over water meadows and canal and drifts slowly through the arch of the brick bridge, from which comes the lazy, peaceful drip of water.

I was waiting one day for two boats to pass through the lock at Newton Harcourt, in the summer. The first sign of their coming was the arrival of the lock-keeper, dusty and breathless on a bicycle, to open the outer sluices and prepare the lock for the crew. He opened two ports on the outside of the lock and the water passed through a subway tunnel, almost without noise, into the lock itself on the inside of the gates.

The next arrival was a tame magpie which belonged to the boat family and flew on ahead of them to the lock. Then came the man of the family, also on a bicycle. A few moments later there was the sound of a petrol engine and the two boats came into sight, one of the old horse-drawn type, towed now by the other, which was a motor boat with a tiller bar of metal piping and a square high cabin painted with broader and simpler patterns than most of those I had seen on other craft.

As they came closer I could see that the house-boat was of an early style, very full of decorations, which had probably been painted at Braunston for the 'main line'. The side panels were pictured and there were flowers above the broad lozenge patterns on all the central gangplank fittings and tarpaulin frames. There was no tarpaulin, but the cords were all in position, some black and some white, making a splendid zig-zag pattern which linked the painted prow with the cabin and the tiller.

Now the boatman opened the main sluices with his lock key and a loud and pleasant roar of water accompanied the filling of the lock. The clear reflection of the sky was lost in a whirlpool of grey-brown water. At the first opening of the sluices the rush of water under the gates was strong and flung bubbles right down the lock, but as the levels became equal there was only a slight surface disturbance.

The mother of the family shut off the engine of the motor boat and her ten-year-old daughter held the tiller of the other. Then the engine was put into reverse as the motor boat nosed its way up to the lock gates; the child, handling her boat with impressive skill, brought it alongside the other, against the solid tree fenders which guarded the lock approach. A second child was playing in the well of the towed boat, swinging from the tie-beams above her head.

Meanwhile the lock-keeper had ridden ahead a few hundred yards to open the sluices of the next lock. On a weekday he would

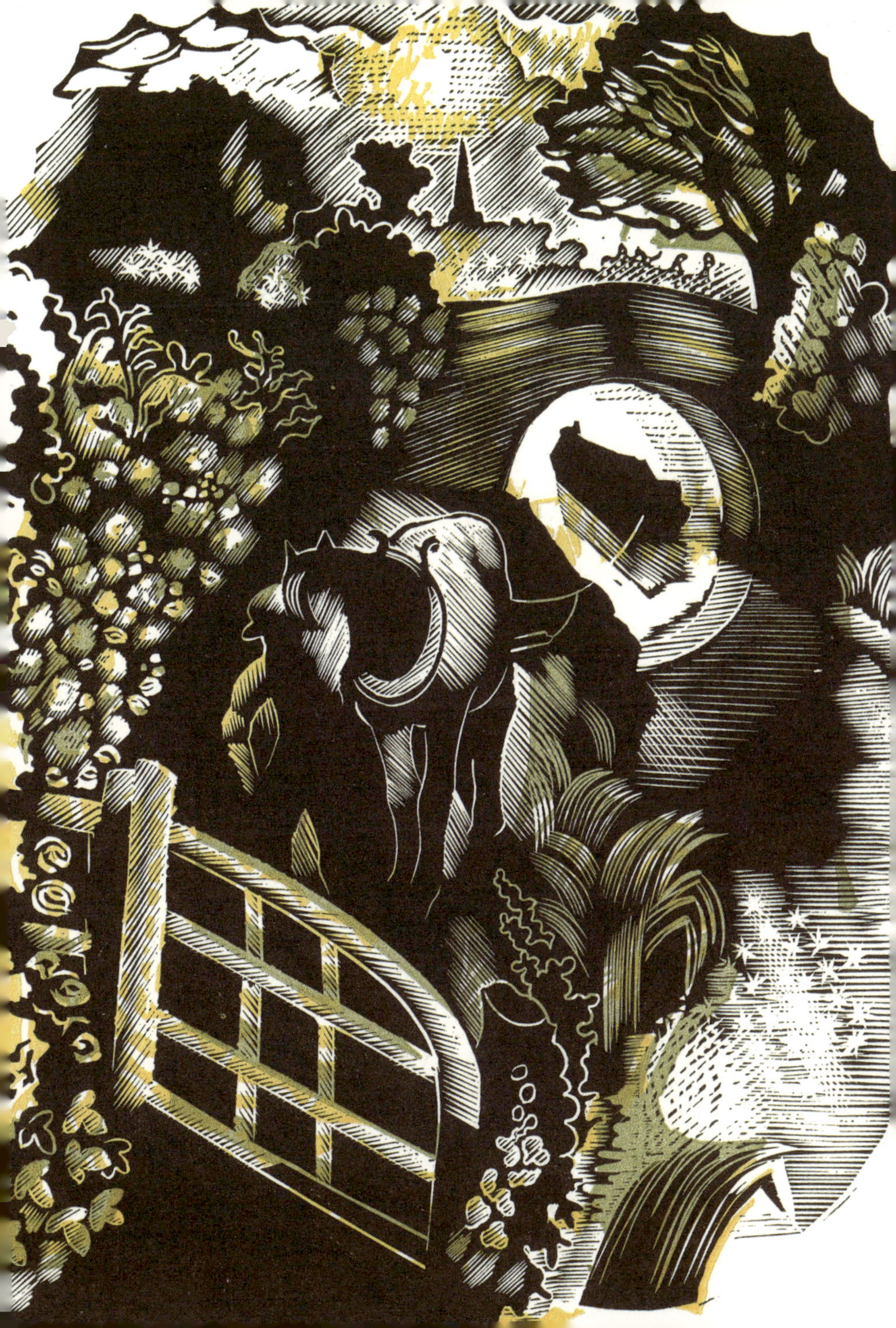

have turned up earlier and filled the lock for the boatman, but it happened to be a Sunday and he did his job with a very bad grace, muttering to himself about losing his afternoon sleep, although the boat people are allowed to use the locks any day of the week so long as the levels permit. He was in many ways an exception to the usual type of lock-keeper, for they are on the whole an obliging and hard-working race. His house and garden were badly neglected. Where some years earlier I had seen a fine display of garden flowers there was a wilderness of high nettles, docks and dog-weed. One small cultivated patch hardly bigger than a kitchen table had been cleared in the midst of bracken, columbine and nettles.

One member of the boat family that afternoon took not the slightest interest in what was going on outside. That was the youngest, Margaret, who stayed in the cabin attacking fruit pudding and a liberal helping of custard with all the energy of a three-year-old. The fair-haired child in her dirty white pinafore and blue frock seemed very remote from the world outside of deep water and hard work. Yet the gay interior of the cabin seemed related to, and a completion of, the wooded waterway.

The highlights of monkey-boat decoration are the castles on the cabin panels, which are rather quiet in tone, the brighter colours being reserved for the lozenge patterns outside. On this boat the castle had a Gothic door and round battlemented towers, but the windows and spires were of a dream world. There are similar windows in some early Victorian gate houses, but the spires, swelling into exuberant gilded onions, have no parallel in England and suggest Balkan or Slav influence. In the foreground a little sailing boat passed under a bridge into a lake beyond, where another boat with white sails headed straight into a stiff breeze.

The trees were bowed before the wind; beyond were snow-capped mountains against a simple blue sky full of white clouds.

The lower parts of the panels were filled with the same type of decoration as has been used for centuries on the walls of houses in Central Europe and linen chests in Brittany and the West of England.

On the stove was a china tea-pot decorated with pictures of a seaside town. The plates in use were plain white, but the show plates which hung by the stove had pictures of flowers or animals and china lacework round the edges. Some of these seemed to be old, for they were more yellow in background than the others, but similar plates are still being made in Coventry, Stafford, Derby and Birmingham.

The mid-day meal had been a disturbed one for this family, as there were twelve locks to negotiate in four miles. I cannot think when the mother had found time to cook the pudding that Margaret enjoyed so much. Dinner prepared in Market Harborough would be eaten on the quieter straights or during delays at locks anywhere between that town and Leicester. Plates would be scarcely dry from one meal before they were taken out from the little cupboard shelves for the next.

When the two boats were in the lock the oldest girl jumped out to close the gate which, even at level water, was almost too much for her. She and I closed one side, her mother the other. The whole business of lock management and eating had been going on simultaneously with this family for an hour or more. This is typical of most of their life, when apparent essentials are carried out at snatched moments.

Lock working is done resignedly by boatmen, with silent, smooth,

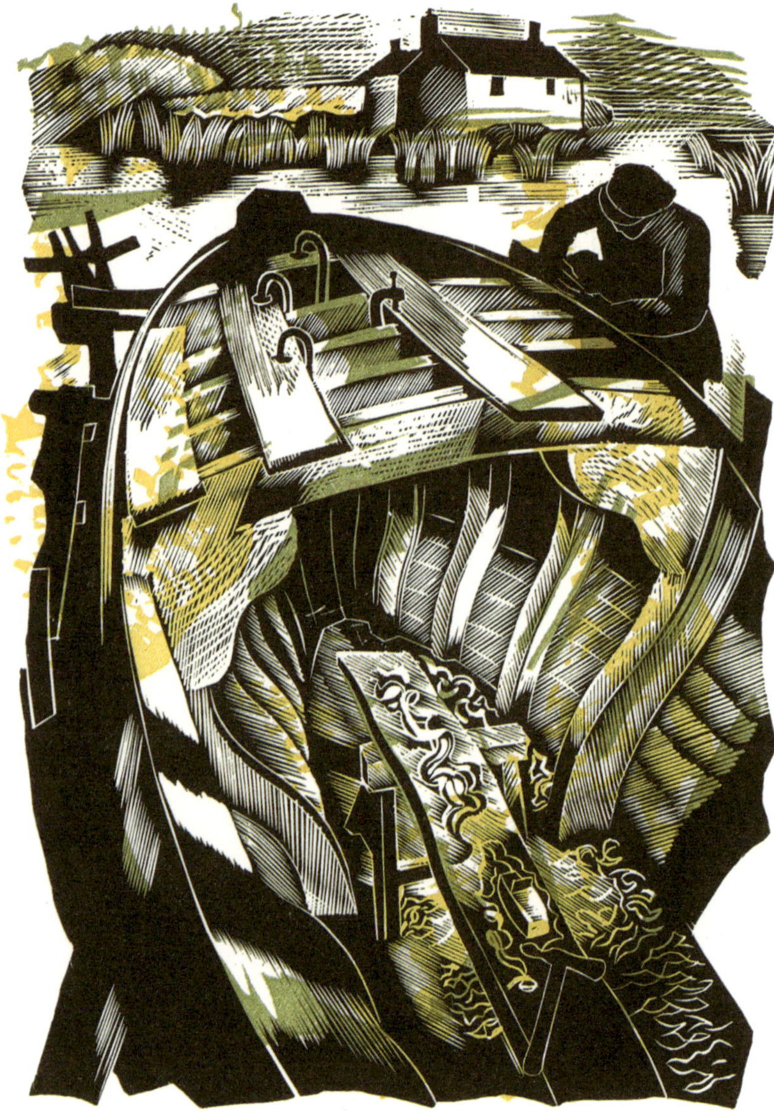

almost mechanical movements and no unnecessary walking or shouting, so as to save effort. When the boats have passed the only noise one remembers is the roar of the sluices, which is a lovely sound.

The motor started up again and the two boats moved smoothly away to the next lock. From Nottingham they would go to the potteries and then back with their cargo in about ten days' time to London. I left them there and went back along the towpath a few miles to Foxton, where there is one of the most fascinating ladders of locks in England.

VI

THE FOXTON LOCKS

Foxton is a scattered pleasant village about six miles from Market Harborough. It stands on the side of a hill, looking out over a wooded landscape into the tree-bordered and intensely green fields of the Leicestershire grazing and hunting country. The canals cut the village into two sections, leaving the church on what is almost an island connected with the rest by bridges.

I arrived there just as a thunderstorm was beginning. Hardly had I reached the towpath when torrential rain came down. This part of the canal, in the village and about a mile from the locks, was overgrown and full of weeds; a thin line of clear dark water, no wider than a canoe, showed where a timber boat had recently passed. The heavy rain woke up the green sleepy surface of the water and shook the dense rushes.

When it had cleared a little, I walked on until I came in sight of a dull grey-painted monkey boat which had been converted into a dredger. Mounted on it were a small crane and an iron bucket, operated by surely two of the oldest employees of the Company. One old man wore the traditional nineteenth-century black bushy moustache of the county, gloriously whitened by age and well in keeping with his hedger's corduroys and knitted waistcoat. These two had known the days when every boat had to be 'legged' painfully through each tunnel by sheer hard work; but when, on the other hand, the ten locks at Foxton could be mounted in as many minutes instead of the tedious hour it now took. This was because of the famous Foxton lift, now dismantled.

The dredging boat was moored in the wide turnabout at the foot of the ladder of locks. On one bank of the approach was the foreman's house, which hardly seemed to belong to the wide fields and thick hedges of Leicestershire, for its upper storey was protected by painted boarding in the Kent or Sussex style. On the other bank, connected by a wooden footbridge, was a smaller house, colour-washed and

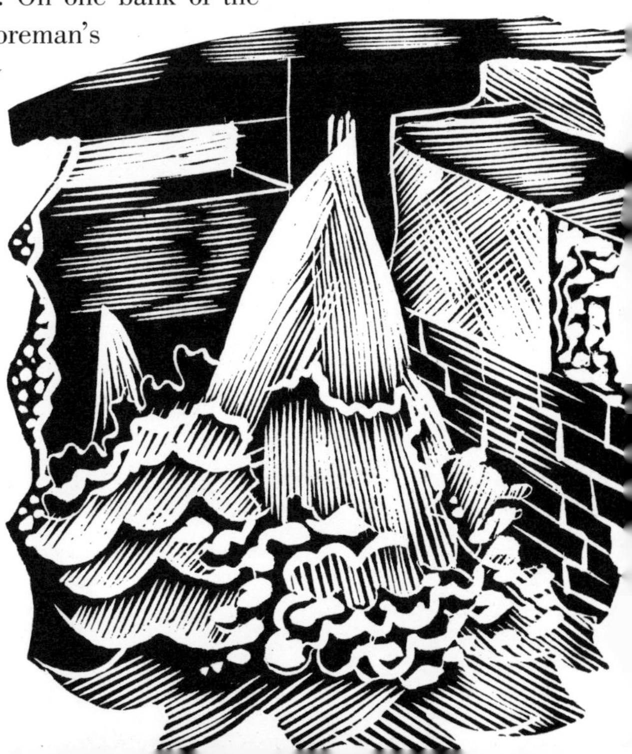

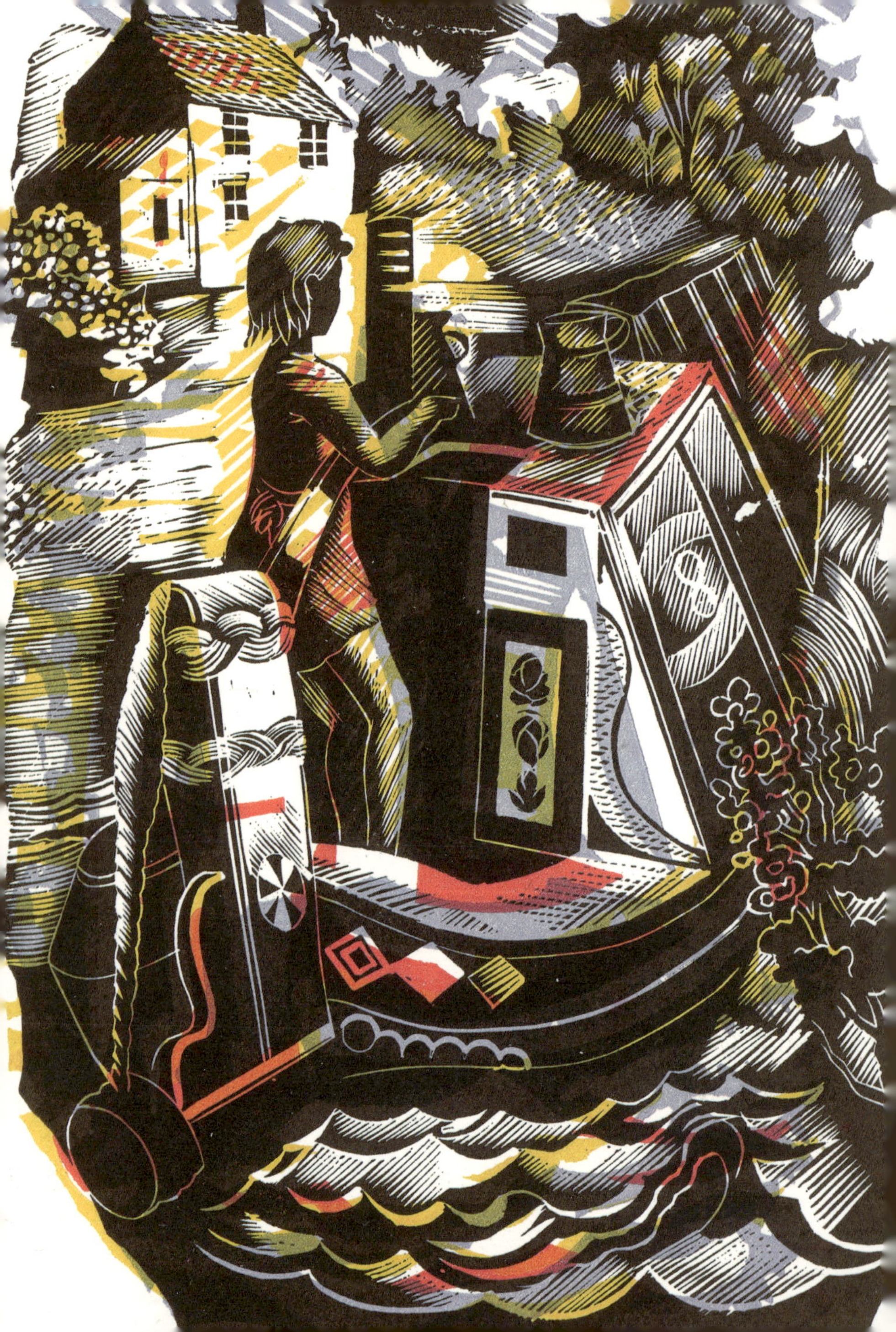

very neat, for the keeper of the lower lock. In a similar house at the top of the lock ladder, about 80 feet up the hill, lived George, who had lost a leg in an accident at the lift. I went to call on him to ask him about the remains of the lift.

'They've taken it all away now,' he said. 'That's where the power house used to be, yonder – I used to keep a bright wheel there when I was younger. There's a bit of the chimney left over there, but it's no good to anyone.'

Deciding to explore, I crossed the lock from George's house to a wild piece of weedy land, crowned with bushes where the mechanical lift had operated. The system was designed on the principle of the twin lifts on cliff faces at seaside towns. At the head of the ladder a small limb of the canal had led into a basin where at one side a metal tank with sluice-gates waited to take the boat. This tank, which was on wheels, stood at the top of a runway inclined at about 40 degrees, with a similar tank in a little dock at the bottom. A boat was floated into each tank and the top one began its downward journey, thus causing the other, with the assistance of comparatively little power, to rise. Then the drop sluices were opened and out came the boats, a hundred feet higher or lower than where they had started. Necessary power was provided by a 250-horse-power steam engine, attached to a little boiler house. From engravings and a photograph I have seen I gather it must have resembled that pride of many a boy, a toy stationary steam engine, even to the neat redbrick chimney.

The whole thing was closed in 1911, chiefly because it proved too expensive. 'For several years now,' George told me, 'wheels, engine parts, bricks, nails and chains have been disappearing.' Today the only monument to an ingenious engineer is a dock which has become a backwater and a hillside so overgrown with pine trees that no one could guess that it was once crowned with strange mechanical shapes.

Standing on the little railed bridge at the top of the Foxton flight, looking out over the Leicestershire landscape, I felt suddenly that I was looking into an old Midland sporting print. Billiant green fields and tall slim elms with bushy hedges in rectangular pattern made a perfect setting for the pink coats of the hunt. To the right, the Langton villages, Church Langton of the famous Adam style vicarage, once home of the enthusiastic rector who filled the fine stone church with brass, woodwind and large choir for a full performance of the Messiah, while large pies baked in the rectory kitchen for the huge party of musicians and singers who had gathered for this early nineteenth century festival.

This is a blackberry valley. If a boat spends any time at a lock lift or awaiting a companion boat at the lower basin, the children are sent into the fields, where the huge Leicestershire hedges are black with fruit in the late summer and September. Deep wicker baskets, each holding about twenty pounds of berries, are taken to dealers in the market town or carried on to London in the season.

It is in this lesser known part of England that the finest hedge cutting and trimming is to be seen. These neatly turned blackthorn and bramble hedges form a solid barrier for sheep and beast and still retain to the full a wonderful bloom of wild roses, elder and honeysuckle above the greenest, lushest grassland outside Ireland.

This hedge decoration seems to form an excellent complement to the white, red and green colouring of the local monkey boats. At the time of the building of the canal these Midland counties were very isolated, and it is a little difficult to realize that this grazing area was once entirely self-supporting.

By this time two boats had approached the lock and had come to a standstill in the turnabout. I went down to see them and found they were my old friend the *Neston* and her companion boat, heading north. A south-bound boat had already started to climb the ladder, so that the *Neston* crew had to wait until it reached the half-way basin, when they would take their boats to the same point.

After about a quarter of an hour Joe opened the sluices and the top lock began to fill. The locks at Foxton are single, so the motor boat went first and the *Neston*, still attached by her tow-rope, followed one lock behind. Sometimes a horse is kept at the locks to drag the second craft after the motor boat. Otherwise the boatman does it by poling or by hauling with a rope across his shoulder. Towing by a chest rope was quite common before the days of motor boats; five or six men would tow even heavy barges for short distances to get them into position. Poling is the easier method: the boatman leans forward with the pommel of the pole against his shoulder and walks the length of the boat from bows to stern, pushing it away from under his feet, by one boat length at a time.

It took nearly twenty minutes to get through the first five locks into the basin, where the two boats were moved over to one side. The family coming from the opposite direction were already there, and a neighbourly argument began. The south-bound boat was shortly due to make a new journey to a place where the *Neston* crew had been several times and the opportunity was snatched to compare notes.

George told me that though canal folk work to no close timetable two boats often met in the halfway basin, which was a great place

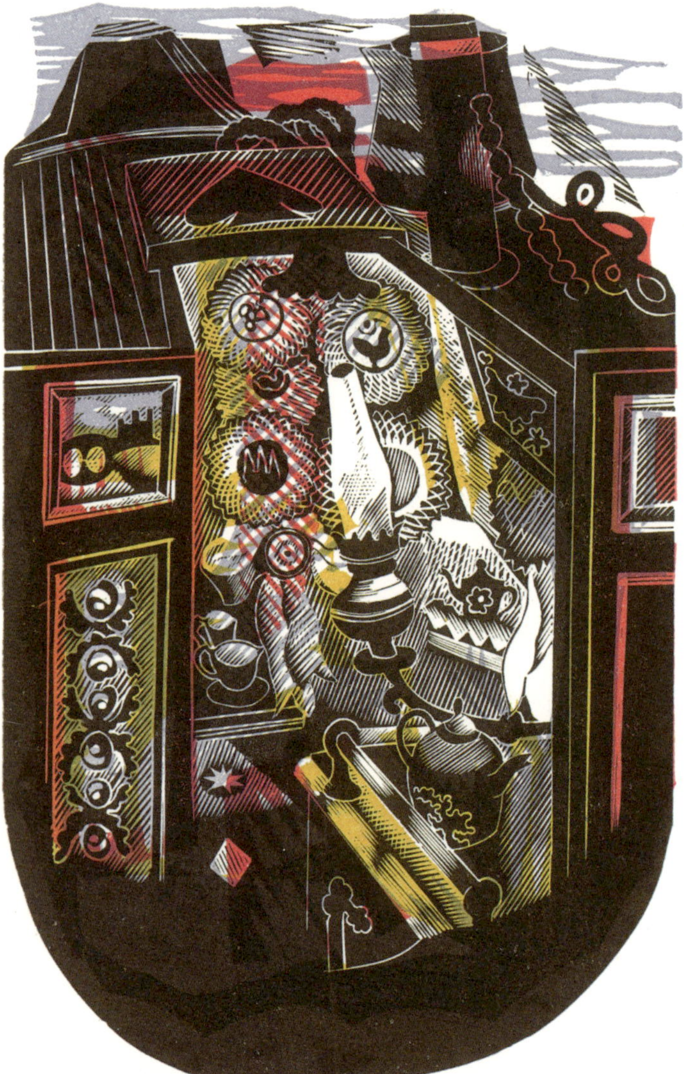

for family reunions; cousin would meet cousin and aunts their nieces and nephews, often quite by chance. At these times all the families lapse into the same broad form of speech, partly Cockney, partly Birmingham and full of lazy abbreviations. The language, though sometimes rich in invective, is not of the sort popularly attributed to 'bargees'. But then, of course, the narrow boat families are not bargees, but boatmen. Barges are at least fifteen feet in the beam while monkey boats are barely seven. The bargee is the more casually employed lighterman or river-carrier from the Thames Estuary or dockland; his language might, at times, be called expressive.

The family in the south-bound boat – mother, father and a little girl called Sheila – were making for Birmingham by way of Braunston and the sleepy Warwickshire country. Then, five or six days later, they would arrive at Brentford to pick up a fresh load; the mother would go shopping and Sheila would go to school.

VII

SHEILA GOES TO SCHOOL

The boat school at Brentford, once a church school, stands on the bank of the canalized River Brent, by 'the Ham'. It is a clean and pleasant single-storey building with large windows and a huge stove. In the winter, when the Thames is high and the storm sewers are slow, it is liable to flooding, so only unimportant things are kept on the bottom shelves of cupboards.

I called there to collect Sheila one morning. At first sight the classroom looked like any other school; but soon I noticed certain oddities about it. For one thing the children, who were all in one class, were aged anything between five and twelve, and there were very few of them. Boat children gather their scrappy schooling when and where their boat docks for a time, which means that the school attendance sheet is highly erratic. There may be twelve children present when school begins, six by eleven o'clock and nine for afternoon school, as the boats arrive at the dock or move off. Sheila was having a whole day at school that day, the first for three weeks.

Along one side of the room were playboxes full of numbered bricks, lumps of plasticine and coloured pictures. On the blackboard was a lesson representing the three Rs all at once which apparently stayed there almost permanently. To the left were set out some letter shapes and two or three lines of writing, in the centre a few addition sums, and on the right a number of words which would improve the children's vocabularies – if they remembered them. In a school where the children come for such short periods, with such long gaps between, this constant lesson on the board is a good way to ensure that something, however elementary, is impressed on their memories.

In an ordinary school one sees on the walls cut-out paper friezes and drawings of aeroplanes, railway engines or houses with gardens and smoking chimneys. Here there was a modest row of pictures – a drawing, done from memory by a boy, of his own boat door, a painting of a castle with towers against the blue sky, and patterns of diamonds in primary colours, some arranged in circular shapes like the roundels and lozenges of the painted boats. In the cupboard were iron holders of wool or raffia with rather spidery embroidery in unmistakably monkey-boat colours and patterns.

Much that ordinary children enjoy in school – dancing, music, singing – the boat children are debarred from, because there are so few of them and they have so little time for the essential lessons. When writing lessons mean only a few strokes at each visit it takes a long time before they can write their names. Reading is easier to carry on in short periods, and is supplemented by scripture lessons, which are held by the chaplain from the Boatman's

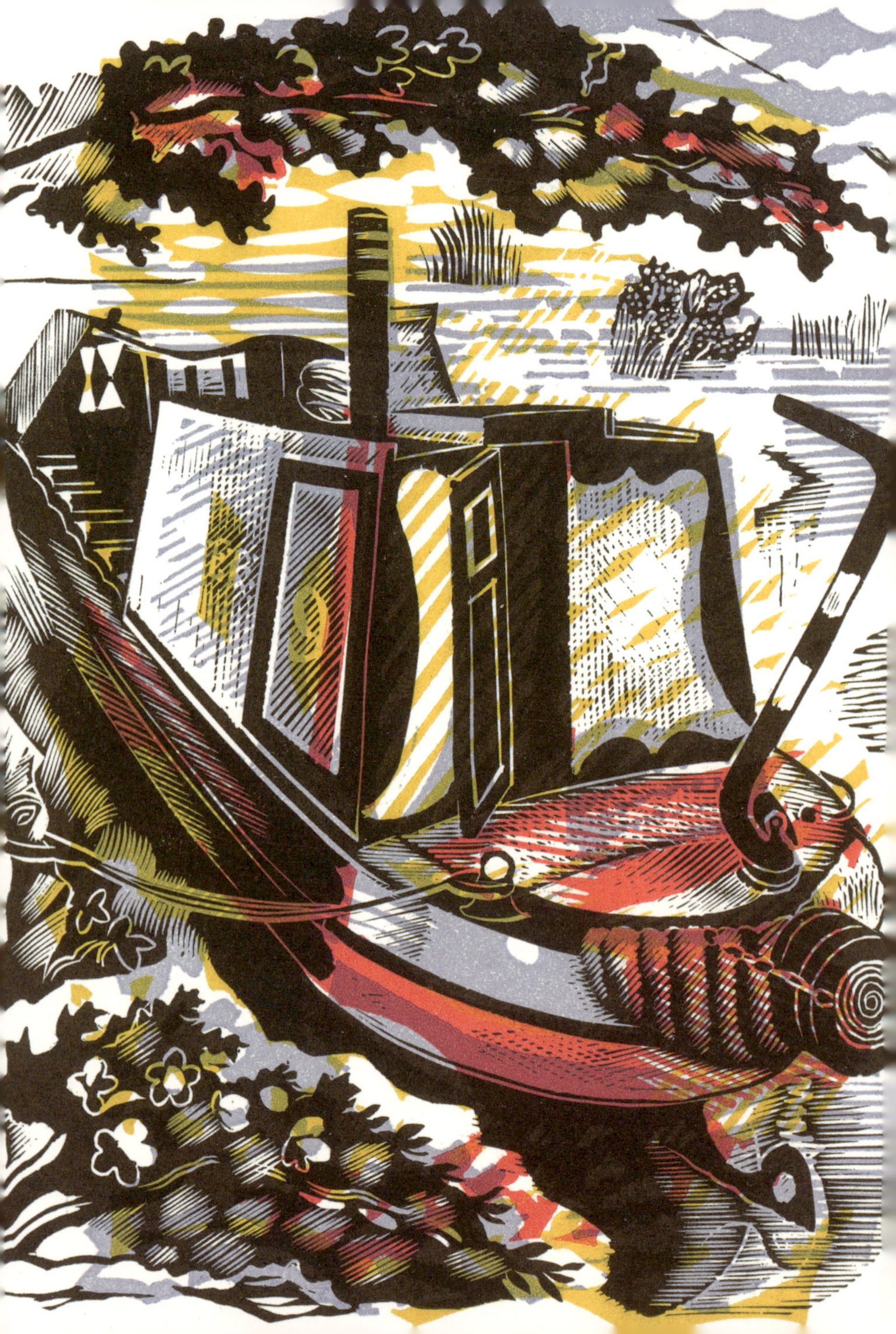

Institute at Brentford. Some, but by no means all, of the boats have wireless sets, from which the children pick up useful general knowledge. The one advantage they have over land children is the extensive knowledge of their own country which they acquire on their journeys.

But to return to Sheila. She has long, very fair hair, big grey-blue eyes, and a fresh bright face. That day she was wearing a little white blouse and grey skirt, and a pink bow in her hair to celebrate her eighth birthday. 'My real name is Sheila Eunice,' she confided to me, 'but I can't spell Eunice, so I don't use it.' Her teacher overheard this and wrote it out on the board for her. Boat family names are usually of the English country type, reminiscent of Thomas Hardy; the older men have names like Utah, Uriah and Ephraim. Many boat families come from Birmingham and the Midland names of Henry, George or James are common.

As it was Sheila's birthday she was allowed to leave early that morning and take me to see her parents who were loading at the dock. Once buttoned into her neat dark cherry coloured overcoat Sheila, in her white socks and red sandals, would have passed for any little town girl. I was surprised that her coat had kept so clean and uncrumpled in the clothes cupboard of the boat. 'Mother hangs it up in the cupboard,' she explained. 'You see, it's not as long as hers. Hers won't go in . . . well, it will, but you can't close the door.'

Sheila was not very happy in traffic and the crossing of the Brentford High Road was a hazardous experience for her. But once we got on to the towpath she was at home and full of information. She explained to me the four distinct types of monkey boat to be seen in the basin and took great pains to point out that the clumsy lighters, which carried produce loaded from monkey boats down the Thames to the London docks, were 'barges' not 'boats', and that her home was a 'boat', not a 'barge'. She took particular interest in the names of boats. There has been so much shuffling and re-shuffling in boat-ownership that these names, often in pairs, show strange incongruities, like *Carp and Iris*, *Gladys and Lapwing* and *Sweden and Buffalo*.

'I'll have to go aboard first,' said Sheila, 'and tie up Peter, because he bites people's trousers if he don't know them.' As a matter of fact, that staunch guardian of the boat was fast asleep under a bunk and never woke up all the time I was there.

'What are you doing here, instead of at school?' asked her mother. I explained while Sheila, feeling overdressed in her white socks, took them off. Her mother was very keen that Sheila should go to school whenever possible. But at Birmingham, for instance, although so many boats dock every day, there is no boat school. All the children prefer their own schools where they can come and go without comment from the others. 'At an ordinary school,' Sheila said, 'they pull faces at us sometimes and call out "boatie".' The town children naturally resent these young strangers whose life seems so exciting and free-and-easy compared with theirs.

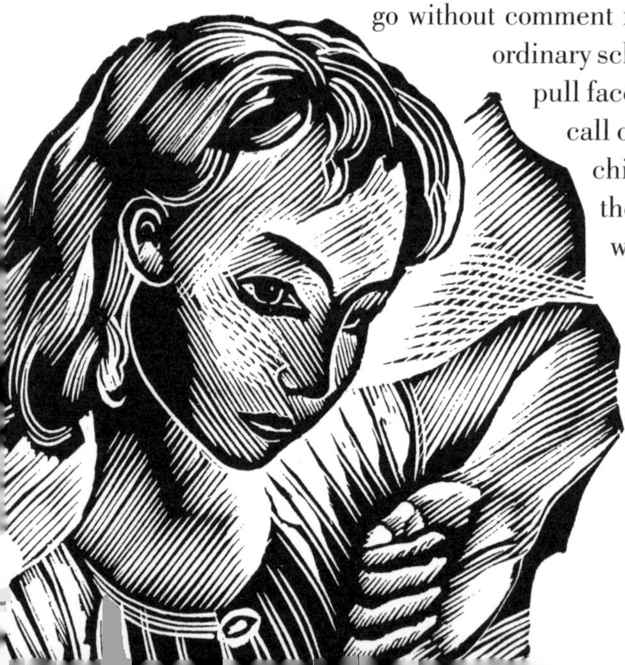

The family had docked at 4.30 that morning and

were already almost reloaded. They would be ready to move off in the evening or at daybreak the following morning. The mother had got in all the shopping, which was a relief to her. The boat families move about too much to become favoured customers with local tradesmen, and often have to put up with the left-overs which the regular shoppers have refused. 'It makes you right empty, you know, of a cold morning,' said Sheila's father.

Their cargo was a strange one for boat-dwellers to carry, for it consisted of sections of pre-fabricated houses. They all greatly admired these little houses because they had electric cookers, heaters and neat little bathrooms. All the same, I don't think any of them would really have exchanged their floating home and independence for a little house rooted in some dull suburb, though it would have been warmer. Sheila's father said, 'If the stove dies down your feet don't half get cold in the boat.' The family confessed that it needed considerable courage to leave the warmth of the cabin for some outside duty.

Their next excursion into the Midlands would take about three weeks. It was to Birmingham and beyond, but they expected to stay with relations on land at Leighton Buzzard on the way, as in Birmingham the boats were both overdue for a refit and would go into dock shortly.

'I wish we could have flowers and pictures on the side of the cabin,' said Sheila. But these boats belonged to a company which no longer kept a painter and so they were doomed to remain in panels of plain red, blue and white. In fact, the only decorations on the boat were the banded tillers and the brasses inside the cabin. On the way back to school after dinner, Sheila eyed rather wistfully the bright water buckets and the diamonds and lake scenes of four boats moored at the dockside. 'Them's Braunston boats,' she said.

Back at school, we found another small boy had arrived who could only stay for an hour. He was busy putting wooden bricks into a box. This was 'stacking', which all the children enjoyed, for on a smaller scale it was exactly the same as their fathers and older brothers were continually doing with cargoes.

Sheila picked up some white knitting she had been doing. The needles were thin and progress was slow, but she hoped to add a few more rows before it was put away in the cupboard for three weeks. In twenty-four hours she would be somewhere among the quiet villages of Buckinghamshire, and in five or six days she would be sitting in a Birmingham school.

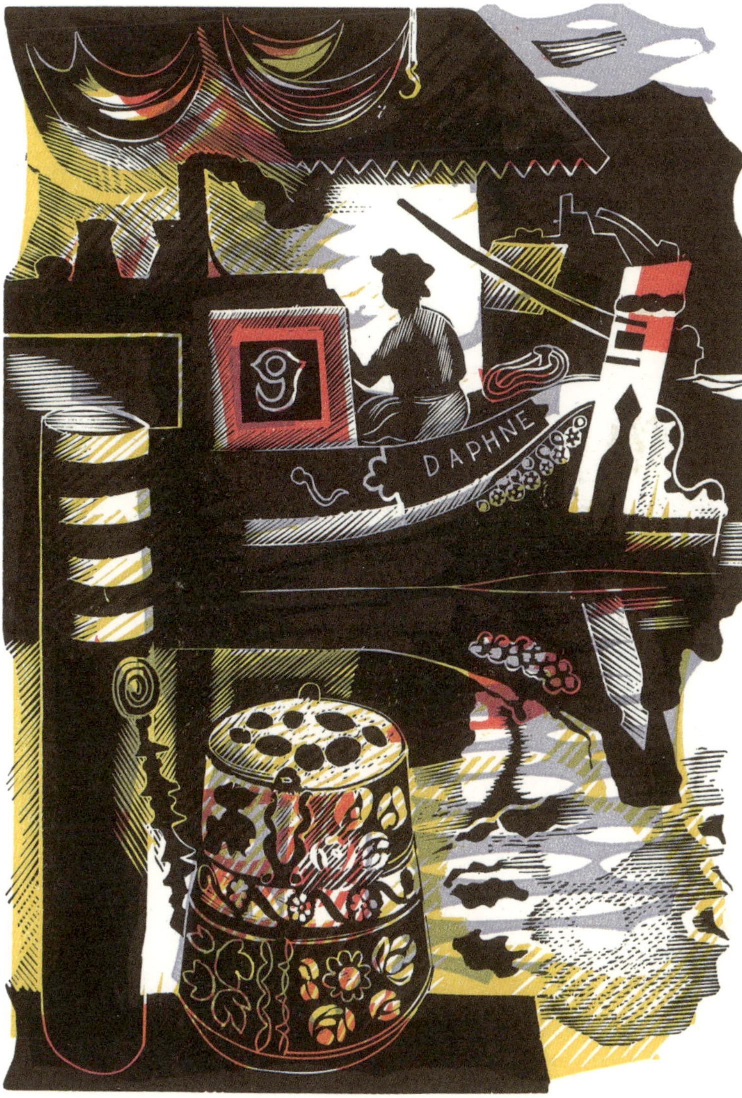
DAPHNE

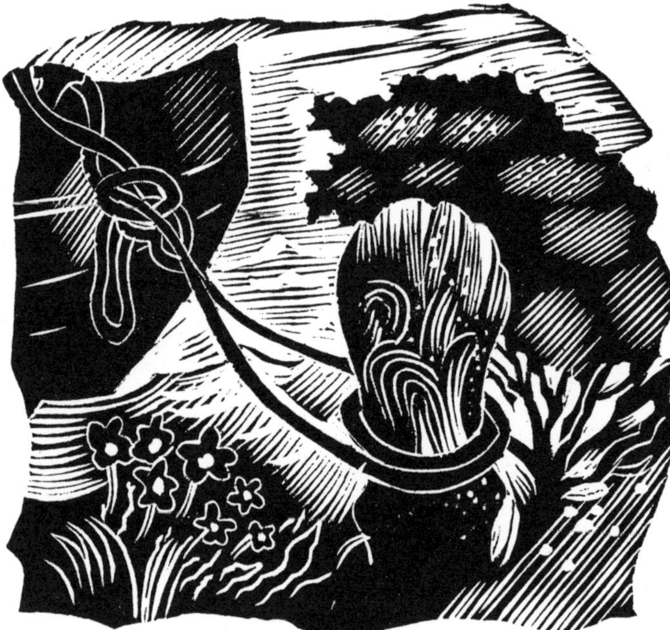

VIII

NIGHT SCENE AT A LOCK

One windy afternoon in October, when the weed flowers and the water marigolds lay on the surface of the canal and leaves and twigs were scattered over the water, I was in a red Canadian canoe a few miles from Rugby. I made fast at a lock-house around which there was a small village, very typical of those through which the Grand Union, Grand Junction and Oxford canals pass. The houses were of the yellow-brown stone or light-coloured brick of the Northamptonshire villages. Elms and sycamores traced a pattern along wide hedged fields and low spinney-crowned hills.

As the sun went down a strong breeze came across the fields, finding a thousand ways through the weave of my tweed jacket. Leaves hurried along the lock side and settled among the reeds at the water's edge. Suddenly the rain came pelting down, and I took shelter in the lock-house. From the doorway, as I went in, I could see a steep stairway disappearing into the heart of the little house; on the right was the living-room, where a fire burned in an elaborate kitchen range.

The light of an oil lamp swinging from the low ceiling shone on the photograph of a fine canal horse, fully harnessed, on one side of the little sideboard. On the other side, scrupulously arranged to face it, was a similar horse held by a youth in a high collar and the Company's cap. The youth, I later learned, was the keeper himself when he worked as a lad at the Brentford stables, now no longer in existence. The horse which had been his great pride was long since dead, but over the fireplace hung the brasses which had adorned its harness.

These were polished to a delicate brilliance which surpassed anything I had ever seen, even in a monkey-boat cabin. There were some fifty or sixty of them, each suspended by a bow of faded silk through holes in the three leather straps which were festooned across the fireplace. Most of them were shaped in the form of a heart, diamond, cone, oakleaf or helm-wheel, beautifully cut by hand and deeply engraved.

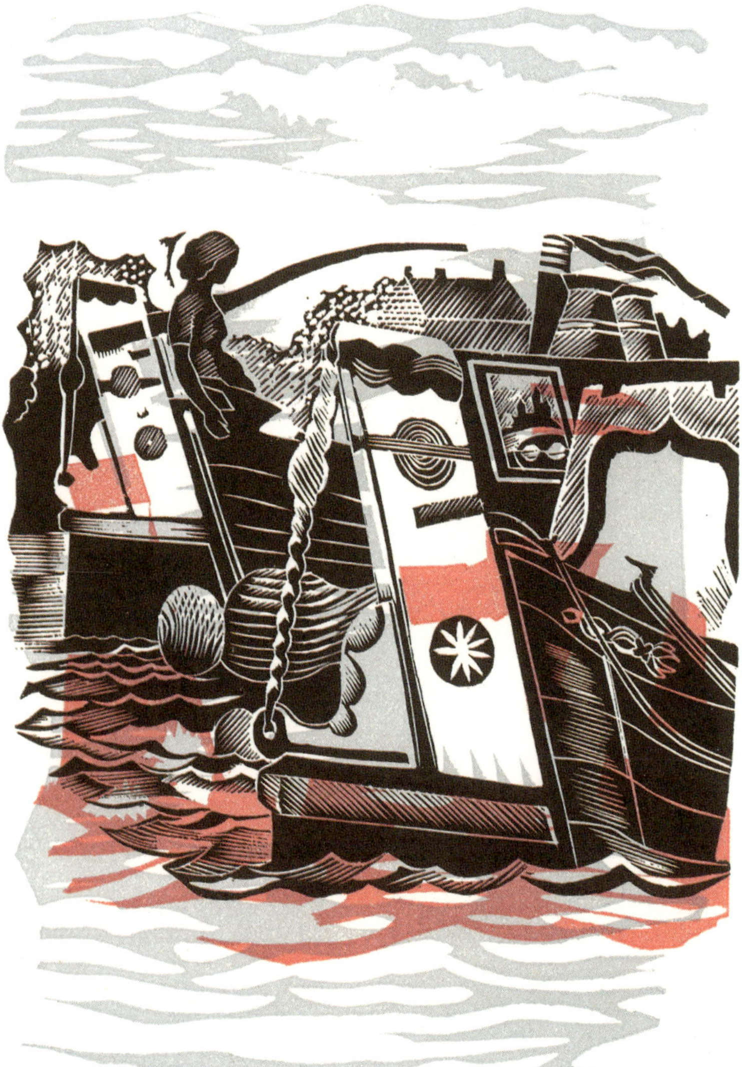

Some were prizes won by canal horses at shows, others represented such milestones as Victoria's two Jubilees and the coronations of Edward VII and George V. Next came some Victory medallions, a Grand Union proficiency badge, and then hearts, diamonds and rosettes to complete the chain.

Sometimes the lock-house families seem a little remote from their environment. One usually catches sight of several children all very much alike in appearance, rather heavily fringed and often frowning. They seem to lead a charmed existence, miraculously avoiding a premature death by drowning. Perhaps because of their semi-private locality, the lock family are often reluctant to talk, promptly retiring behind the panes of their small windows between the passages of the boats. There is usually a dog by his kennel, a leafy row of summer beans, a latticed porch, and almost invariably roses. It is their white palisades, however, which give these little houses such an attractive appearance.

Lock-houses have a double charm. There is the purely domestic aspect of them, but they also have something in common with railway signal cabins, without the mechanical complications. Set midway along the lock, they remind one of tiny country station houses, where the station master's wife can do her cooking and at the same time keep an eye, from the windows, on her husband's branch line.

Just as a signalman knows when a train is due, so a lock-keeper receives good warning of a boat's approach. This seems unnecessary today, for there is a limit to the number of craft a lock can accommodate at one time or in one day, but was a vital precaution in the heyday of canals. Some of the older lock systems lose a good deal of water at each working, so that more must be taken from the feed. For this reason it is customary for two or more boats to pass through the locks together.

A pair of boats was due then, an hour overdue in fact. I thought I saw lights and pressed my forehead to the window, between the lace curtain and the potted plants, to get a clearer view of the darkening landscape through the rain-washed pane. The lock was the first of several after a clear stretch of ten miles and I wondered if the crew would go through, as it was already dusk.

Nearly an hour later, when it was pitch dark, the lights of two lamps, one above the other, came into sight. The lower one shivered, and I realized it was the reflection in the canal of a solitary headlamp. As the boats came nearer the beam glittered through the driving rain, and the fast-forming puddles on the stone edging of the locks shone like gold.

I put on a coat and went out as the first boat slid into place before the lock gates. The strong headlamp lit up the whole scene with a stagy light, in which a young man wearing a long rubber mackintosh hurried into view with the sluice key slung across his shoulder.

When the roar of released water in the blackness of the lock had ceased the two boats nestled side by side against the walls and the sluice began to fill. As the level of the water slowly rose the light of the headlamp rose with it, picking out: first the wet lower gates of the lock and then shining across the water to strike the white gates of a house. At that moment the heavy storm of rain passed over, and a faint moon appeared sulkily between the clouds.

The boatman's wife had opened the half-door of the cabin, for the rain beat from the bows. I could see her busy making a stew, ready for docking time. When the rain stopped she opened the second

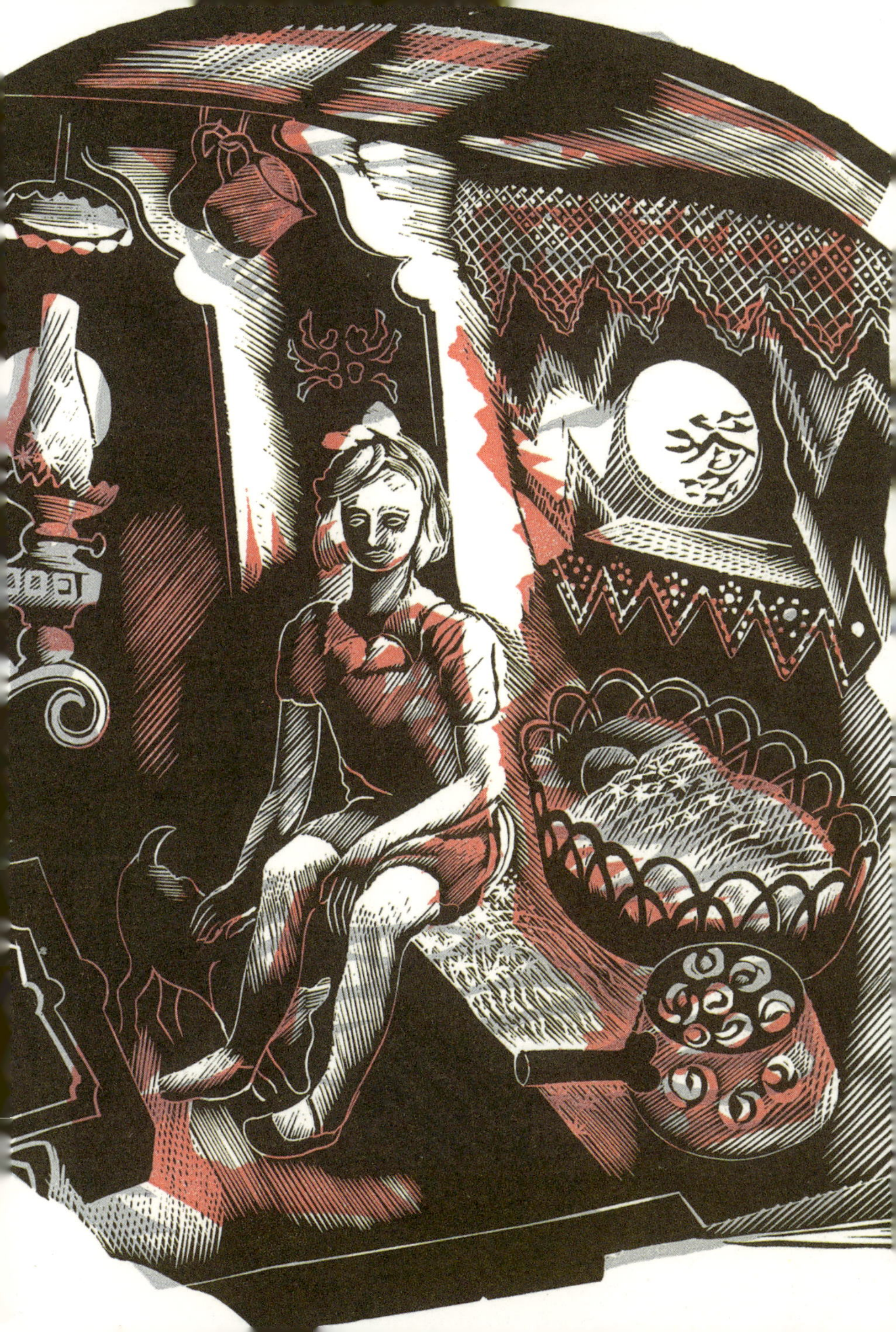

door and smoke from the stove poured out of the cabin. She looked tired and dirty after a hard day. Her apron was a muddy grey and her weather-tanned arms were thrust out of a tight and rather grubby blouse. She belonged to the warm lamplit world of the cabin, which seemed an entirely different world from the wet lockside with the cold blue light of the moon on it.

Water dripped from the willows on to the roof of the cabin and great drops splashed in the water can and on the canvas covering. The cabin was disordered and stuffy with steam from the stove. On the seat was a bundle of clothing and a raincoat thrown over a grey-green eiderdown quilt. The quilt moved a little; a tousled head with big black sleepy eyes appeared for a moment, looked around the cabin and disappeared as the baby settled down to sleep again.

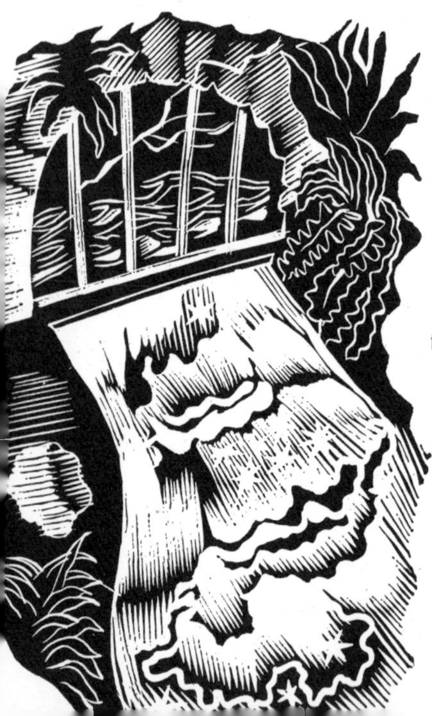

Now the second gate was opened, the engines started up again and the two boats moved off. As they did so the rain came again from the east. The woman, putting a shawl over her head, came out from the cabin and took the helm. I watched the headlamp pick out the trees and yellow reeds along the bank, until the man decided he could see the waterway in the glow of the clouds and turned off the lamp.

IX

REFLECTIONS

A canal in a pastoral setting, with trees bordering its edges and the water full of blue sky, is a very beautiful thing. But a canal set among the mournful industrial towns of the Midlands can give one strange shocks of pleasure; its life and beauty lie in its reflections. When these cannot be seen, when the water is ruffled by wind and rain, it is as though a blind had been drawn down; the canal is dead. I shall never forget some of the reflections I have seen. One was astonishingly dramatic, amid its dismal surroundings.

At one point the canal to Wolverhampton runs very close to a reservoir in the centre of Birmingham. On a wedge of land there showmen of the last century, with great energy and a modicum of taste, had built an amusement park.

When I saw it, almost everything was in a state of decay. Over a wreck of faded plaster, which had been the main entrance, broken electric light bulbs swung from painted arches. Mud and black water, settling under a hut of cheap colours, oozed beneath the fences of the main enclosure. Inside it, large tarpaulin-covered shapes showed where once amusement booths, calliopes and heaving dragons had carried on a busy trade. Now everything stood silent, a picture of neglect.

Here the canal is wide and an old timber jetty, leaning lop-sidedly, is built out into the water. Rushes grow beneath it, the tallest leaves filling the gaps where the floorboards should be. It was from the end of this jetty that I saw the reflection of a circle of wooden horses under a gay roofing.

Each horse was poised in a frozen gallop on a twisted barley-sugar pillar. Some were gilded, others white, dapple-grey or black. All had painted on them flowers and baroque harness, such as circus ponies wear. One horse was spotted all over with the sort of flowers you see painted on the sides of a monkey boat; full, bulbous blooms growing fan wise on spiky stems from a central spray.

In the gleaming dark-green water the colour of the horses was subdued and the gilded shapes looked warm against a leaden sky. A cloud moved away from the sun and the reflected tarpaulins and chimneys glistened in the April light. Looking up, I could see between the horses and the swaying canopy the gilded dome of the Town Hall, trying in vain to make the Midland capital imposing. The whole scene was a strange patch of enchantment in a forgotten backwater.

Another reflection I remember vividly showed the industrial canal in its grimmest mood. The place was a waterway which runs from east to west near the heart of Birmingham. There is a cutting

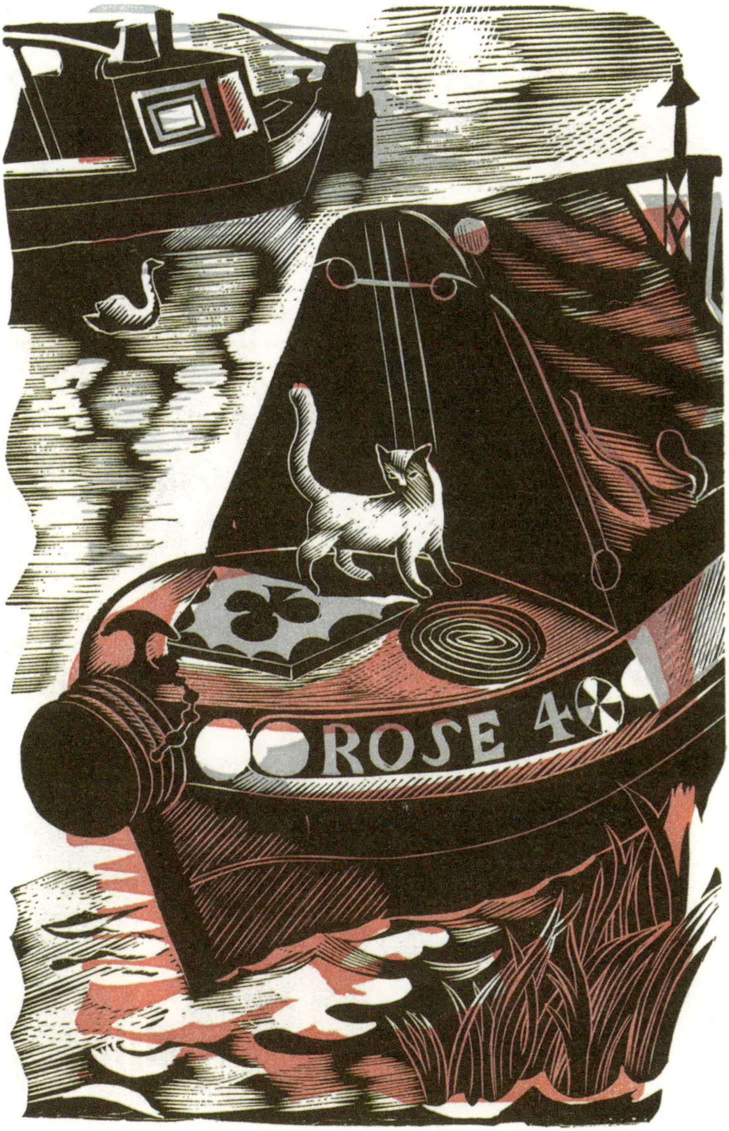
ROSE 4

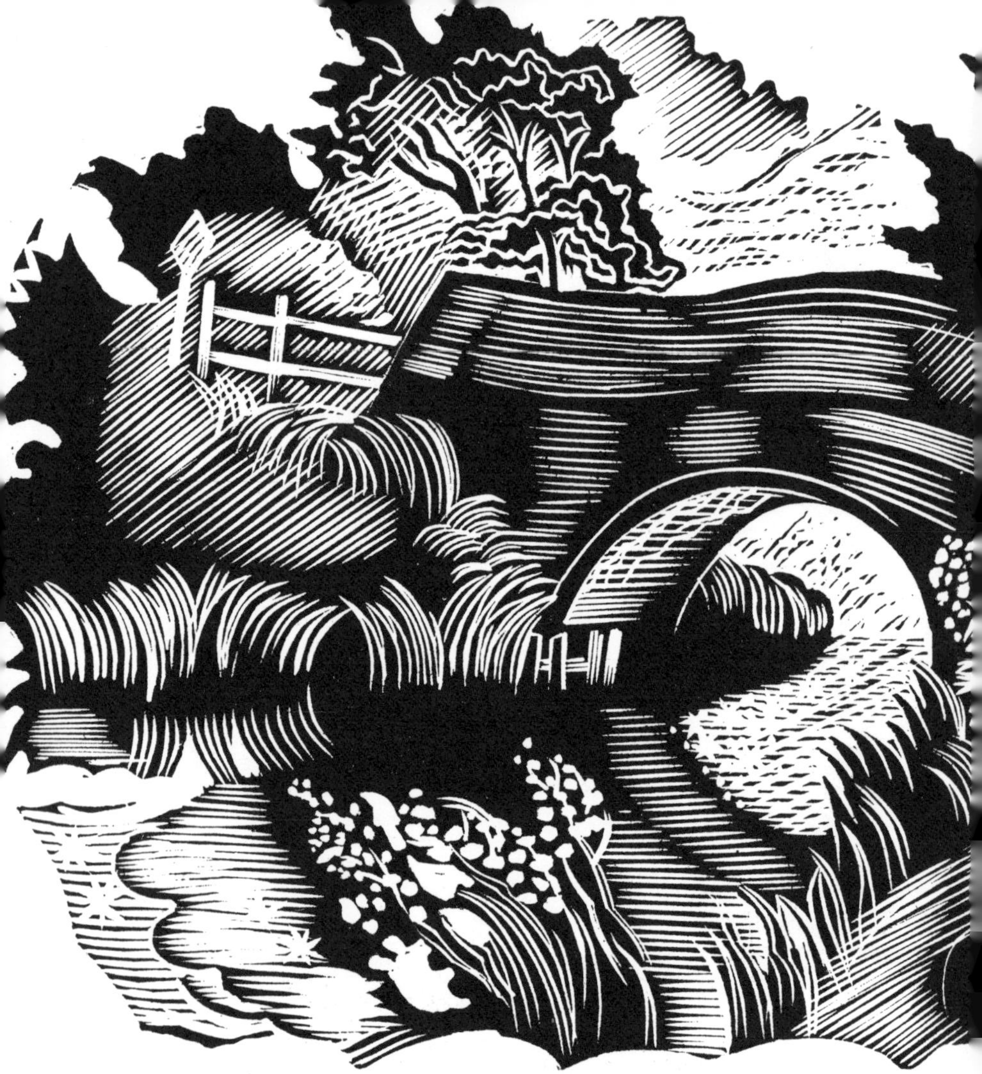

there which is bordered at first by dull red houses. Farther on it develops into a deep, wet, brick-faced excavation, passing down a succession of locks, deeper and deeper into the shadows of tall offices and factories, with dripping walls and trampled towpath. Halfway down this weeping gully, in a puddle of water on the

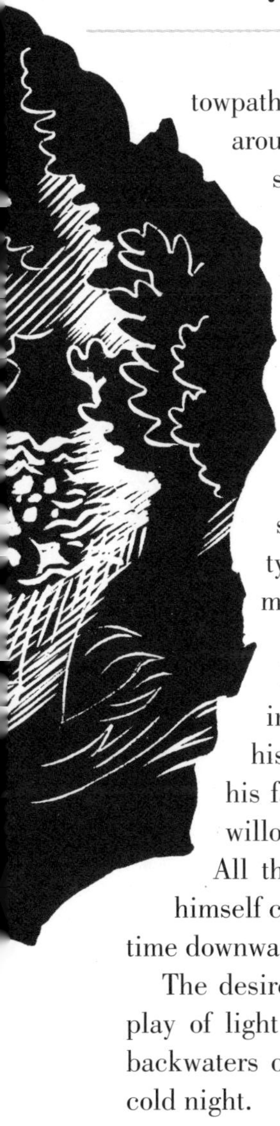

towpath, I saw a brilliant white cloud in blue sky, while all around was dark green and stormy. The bright cloud seemed to hold a promise of the sunny, peaceful, lazy countryside which lay beyond the city.

Another reflection which I associate with the Midlands in sleepy mood was in clear water, which showed a replica of a boat, every detail complete, against a luminous sky. Behind it was the half-circle of a humped bridge with each stone perfectly repeated in the water. Tall elms lay black in the canal behind the bridge. This scene, with a quiet sky of clouds resting on each other, struck me as very typical of the middle counties in their most pleasant mood.

I once saw a boy fishing from the edge of a narrow boat about thirty miles from Uxbridge. As I stared into the water I noticed the oddly surreal character of his reflection. The nails and irons of his boots framed his face, and his head was almost hidden by long heavy willow branches which jutted from the sky towards him. All this was quite clear in the water image, yet the lad himself could not be seen for the screen of leaves pointing this time downwards to touch the water.

The desire to observe strange things from reflections and the play of light on water led me to plan a night journey into the backwaters of a canal, an adventure I later undertook one very cold night.

X

VOYAGE IN THE *EVE*

A friend of mine had a small motor boat, which shared a Leicestershire wharf with a family of ducks. She was a most original craft, having a fishing boat hull about twenty-two feet long and five and a half wide into which had been fitted an ancient, noisy, but efficient gas engine. Atop she had a strange superstructure of a cabin with two sliding windows and a sliding roof, like that of a monkey-boat; a cabin cold and draughty in winter and stuffy and hot in summer. Her colours were mostly those of borrowed or stolen sea-going fishermen's paint, so that her hull showed that clear blue that sea gives to the richest paint, while her cabin sides were gleaming monkey-boat white, with dull grey at the windows and a deeper grey on the roof. But the real glory of the *Eve* was her tiller which, though rarely used, as the rudder had been connected to a motor car wheel in the little deck house, had

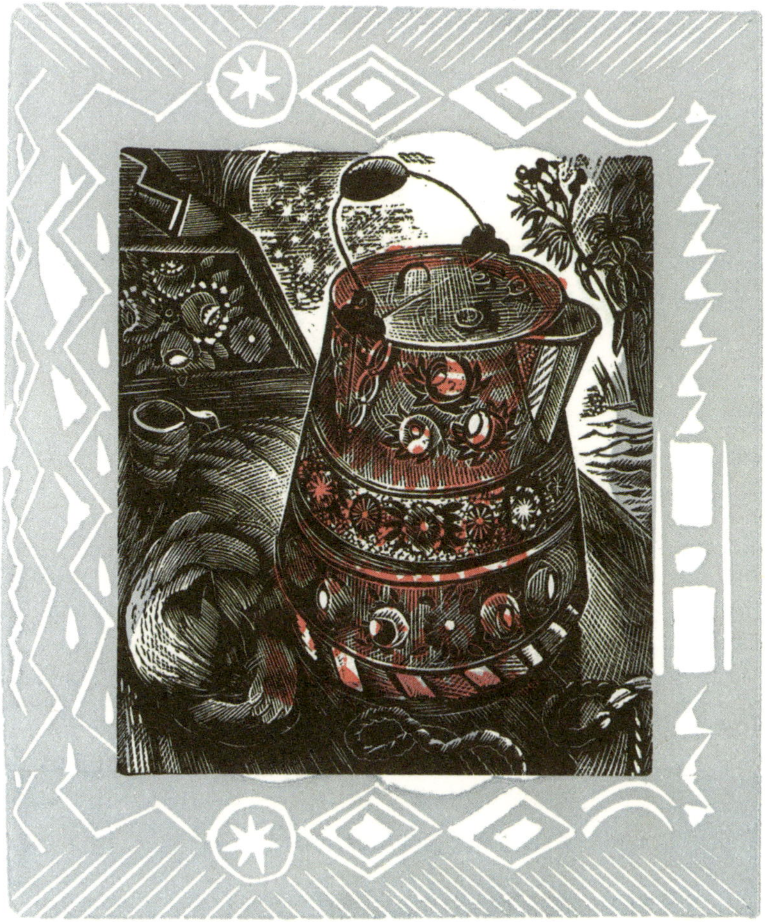

been cut, painted and erected by a monkey-boat craftsman. It was painted with red, gold and white diamonds and bore the traditional white plaited cord. The tiller bar, too, was banded in blue, red and white and surmounted by a knot of laced cord.

We left the wharf on our voyage of observation one February afternoon in a dull yellow light, our gas engine running with a violent rattle. Our route lay along a semi-canalized river between deep-hedged fields, fringed by long decayed grasses which lay untidily in the water. We passed under many bridges, each with a whitened fencing as typical of the county as the deep red houses along the bank roofed with blue slates or rusty-coloured pantiles. At each bridge was a bridge house, most of them being the local fishermen's inns and bearing pleasant angling names like the Pike, the Grayling or the Fish.

We had taken down the tiller and were shut in the cabin, surrounded by the little windows, which had no glass, only shutters. On our way we passed numerous little houses and spinneys and cheerless wet villages from which lights were beginning to show through the greying twilight. The noise of our motor frightened the sheep penned by the waterside and the lonely bark of a dog hailed us from a distant farm. The first to give us friendly greeting was the keeper of the lock-house where the river branches into a canal. It was the backwater of this canal that was our destination. We paused for a chat with him in the little white lock-house, and were given bread and cheese and hot strong tea by his wife. Then back to the *Eve* again, blinded by the lamplight and groping our way until our eyes got used to the darkness and we could see her dim and friendly form.

A little moonlight, filtering through a thin cloud, showed the path of the water, as the gas engine sprang into a hideous din, which later subsided into a pleasant, steady rhythm as the *Eve* nosed out into the waterway, heading for the next bridge. For an hour we slid under bridges and into gloomy tunnels made by giant elms and close-hanging willows until a white fencing told us that the time for more difficult manoeuvring had arrived. We stopped the engine and paddled the *Eve* into the narrow opening of a backwater, under a wooden bridge which cleared her cabin roof by only six inches.

Beyond this opening the waterway grew broader until it was about three times the width of the canal we had left behind. We paddled the *Eve* from stern and bow across this tiny lake towards yet another narrow opening, hardly discernible in the dim light. Our landmark was the gaunt bole of a decayed willow, with wild arms outstretched among the elder bushes. Here was the entrance to a yet wider stretch of water, which turned and twisted away in the moonlight.

Here we stopped paddling and returned to the cabin, lighting the brass-reflectored oil lamp and opening the shutters as the boat drifted slowly forward. Soon we could see some of the wild life we had hoped to see – a moorhen, sitting on her nest in the reeds, and a little farther along a crested grebe nesting, with the male bird moving about among the grasses. The birds took on a new beauty in the yellow light of the lamp, and no less beautiful were the leafless branches of the willows and the silver and black leaves of the evergreens.

Farther along we disturbed water rats, swimming merrily, and a fox scurried anxiously from behind a tree, only to find himself in full floodlight. Rabbits, disturbed by the noise of birds and the movement of water, looked pale and transparent in the lantern-beam. The light played on the frosted willow-stems and fell suddenly into deep dark water along the bank. The leaves of the evergreens, thinly frosted, and the spiders' webs festooned between the rushes, were reflected in the water of the canal, moving a little as the light night breeze played over them.

After a few hours' sleep in the cramped quarters of the *Eve* we stirred at five-thirty to see a grey-yellow misty light over the little lake. As we made coffee, our world grew larger; distant tree tops became clearer beyond the bushes close at hand. Where there had been the night before a solid grey-blue space behind a lighted tree, there was a little landscape with red-roofed Dutch barns and a big brick farmhouse. The yellowish mist gradually gave place to a darker and colder atmosphere, so that the farm had a cheerless look, in spite of the smoke curling from one of its chimneys.

At this hour the rushes are greener and the water seems deeper and very sleepy. Some two or three feet below the surface I could see large leaves, covered with a thin moss, moving very slowly in the current. New spiders' webs had appeared along the side of the cabin and between the tiller bar and the mooring ropes. The parts of these ropes which lay under water, and which had been clean and bare the night before, were

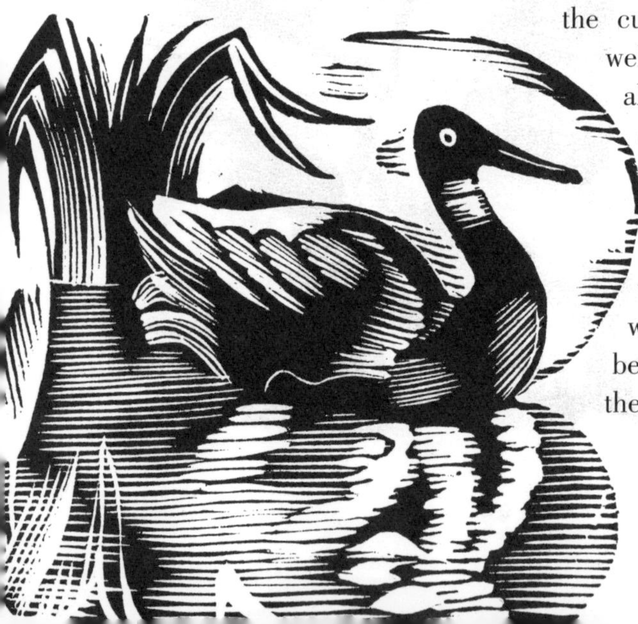

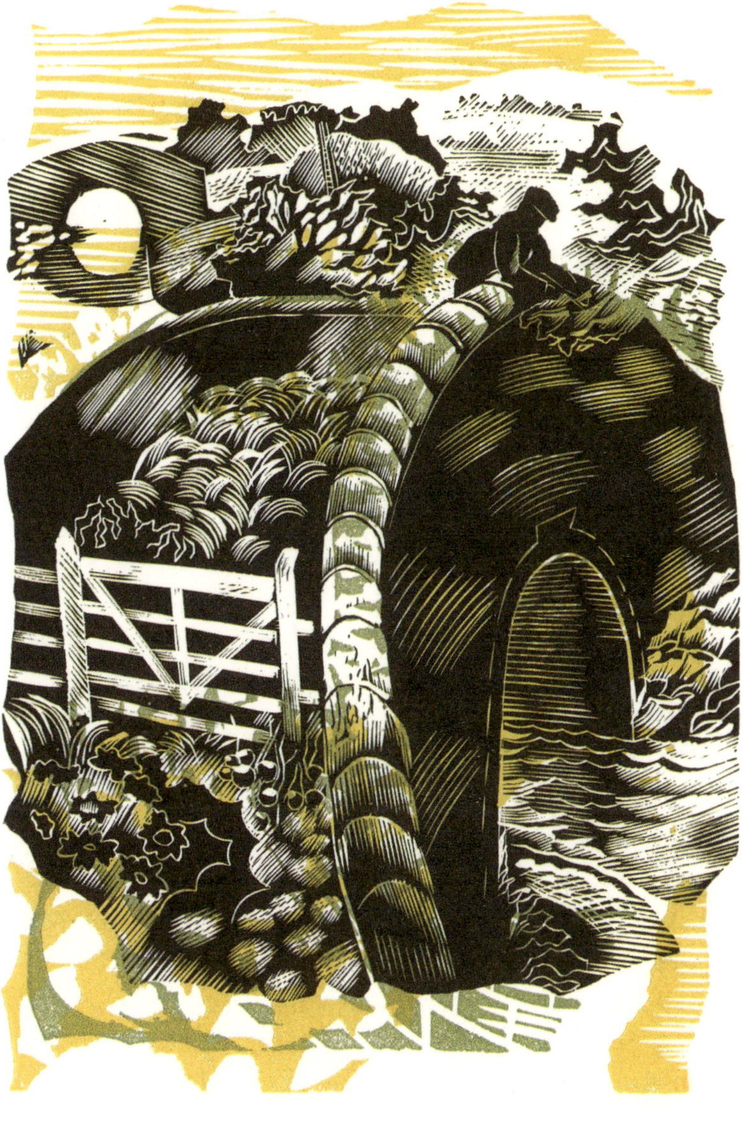

decorated now with pale green threads of weed and moss.

The water was cold and silky to the fingers as we dipped a decorated bowl into it from the stern of the *Eve*. Eyes refreshed and freed from the smoke of the paraffin, we could see beyond the village a tall thin church spire, a low valley side edged with trees, a rough skyline of cottages and a tower where another Fosse village stood. On this morning I envied the people of the narrow boats, who awake each morning to such surroundings.

Soon the sun lit the red wall of the farmhouse and poured over the meadows, where the dew lay heavy on dead thistle tops, ragwort and tufty grass. Cobwebs glistened between plant and ground; the smoke from the chimneys reddened, and all the water was sunlit and warm in colour. We loosed the ropes and paddled silently out of the pool, as two people might leave a church service before it was over, until we reached the opening. Once free of the weeds of the backwater, we broke every spell of morning peace, startled fowl, water rats and pigeons with the clatter of our backfiring gas engine, and rattled ahead to the lock-keeper's house, our minds fixed hungrily on breakfast.

XI

THE ROMAN CANAL

I set out to explore the waterways of East Anglia one day in a motor boat called the *Girl Pat.* Although her hull is very sturdy and heavy, she has a suggestion of sea-going lines about her, for she was originally built in a Hull yard on the Humber estuary. Nowadays she has all the characteristics which unmistakably mark an inland vessel: her washboards are very low, useful only against rain and spray from the wash of other craft; although she has a wheelhouse, it is limited in vision, with nine-inch high windows on three sides, only half of them fitted with glass. She has one mast, a slender affair compared with a sturdy yacht mast, and only a single brown and ochre sail. Her colours are very much those of a canal boat, with yellow-painted deckrail, white wheel-house, blue waterline paint and a bright helm.

As she made her way through the pastureland of the West

Lincolnshire plain a long line of hills appeared to the north. Those same hills had met the eyes of thirteenth-century boatmen, paddling great barges full of Normandy stone along the Dyke from Hull for the building of the great cathedrals of East England and the fine church Bishop Hugh had planned at Lincoln.

Today St Hugh's great presbytery and choir, with three later towers rising from a grey-green lead roof, stands high on the ridge of the scarfed hillside. A stronghold of security and a link with the water-traffic of past centuries, the great bulk of the cathedral, golden-yellow and light green in the sun and shadow, glows above the water of the Foss Dyke, towering over the low farms and the heath and fir plantations. The shepherd still blows his long horn from the pinnacle of the West facade and by his side the Kings look out over the wide land to the spire of St Mary Magdalene at Newark, the almost Rhineland castle of Belvoir and the Gothic lighthouse tower of St Botolph at Boston.

As we drew nearer to the old Roman town of Lincoln, the hill, castle and cathedral became more full of colour. The pattern and structure of stone grew clearer, the pale jade and yellow was mingled with grey-white limestone; against the colour of the walls the black and deep blue of the wet side of the dripstones stood out clearly. The lead roof looked less green, and parts of it were a cool steely colour from repair and recent workings.

The *Girl Pat* passed through a lock into a basin called 'The Pool', which is in fact an inland waterway harbour edged by warehouses, granaries, stores, little shops, a church and railway yards. A waterway running alongside railway yards is just one of the incongruities of a city which has a medieval heart and a modern steel industry on its outskirts. The river craft here – registered at

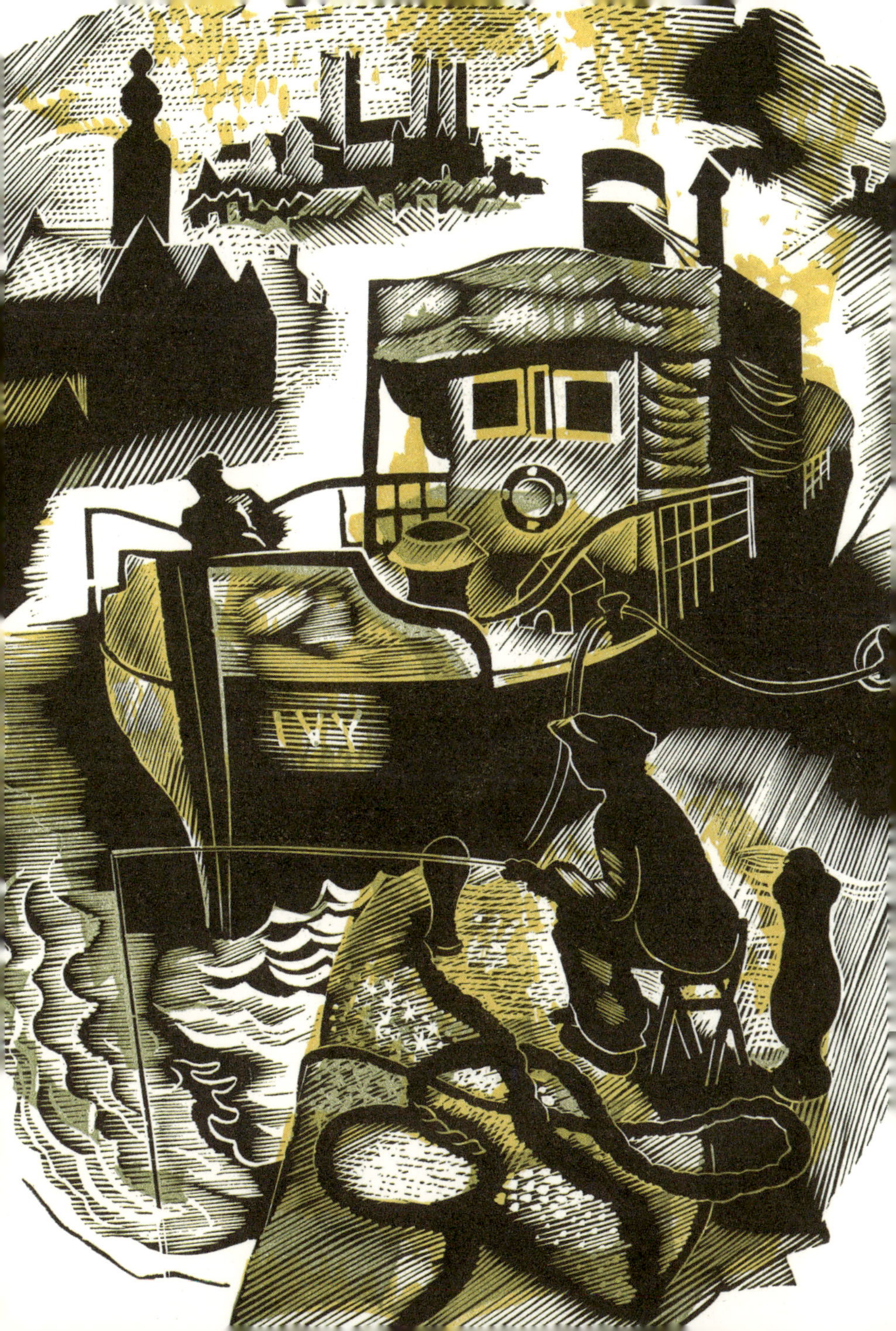
IVY

Hull, Burton-on-Trent, Birmingham, Stafford, Goole and Selby, in Yorkshire – are rougher than the narrow boats of canals farther inland, and have something of the exciting look of ocean-going vessels.

Alongside the quay, the *Girl Pat* sounded her horn and we edged around a large boat to the dockside. Originally a lighter, she was fitted with a derrick by which sacks of grain from Hull were being unloaded into lorries to be taken to the mills. These actually were only about half a mile along the waterside, but the boat was too big to pass under the arch carrying the main road across the canal. Her load was completed with sacks of sugar beet, shipped by water from Kings Lynn. How these arrived at the Lincoln docks I cannot imagine. Inland cargoes have a habit of turning up in the most unexpected places.

Out of the basin a swing bridge leads to the canal which runs through the centre of the city. This has a notice on it: 'Creels, lighters, barges and other craft entering the canal section must not allow anchors to drag.' It seems strange to speak of anchors and not mooring ropes on a canal, but then a good deal of the Foss Dyke and Wytham shipping is from estuary to sea – from Hull to Boston, for instance, or from the Humber harbours to the Fens.

We passed through the swing bridge into the canalized River Wytham, lined with dull red mellowed buildings. A tall narrow granary, built in 1740, is now a Boy Scout headquarters, outside which an almost unbelievable spiral staircase, uprooted from some seaside pier, has been made to serve as a fire-escape. The canal is bordered with wooden post-and-rail fences and over it arch at regular intervals Chinese-looking and almost semicircular metal bridges.

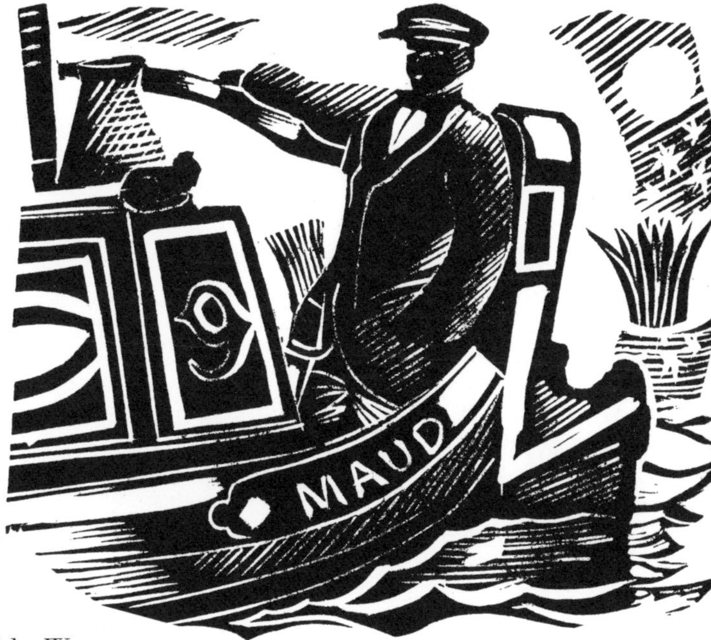

Farther on, the old and elaborate hoist of a granary and bakery leans out over the water, supported by a decorated iron frame of complicated design.

The inevitable Waterway Inn, a tall yellow building, stands head and shoulders above the small houses surrounding it. Postmen, delivering letters, trot backwards and forwards over the little bridges to the houses on either bank. Each bridge is different and bears its own gas light in Edwardian splendour.

The *Girl Pat* was going on, through locks and sleepy windmill villages, but I returned to 'The Pool' to have another look at the boats there. One, the *Ivy*, which used to travel between Lincoln, other inland towns and Hull, has been converted into a houseboat in which a whole family lives and works. I would call her a steamer, except that probably she no longer has a boiler in her. She has a long black and red funnel and a tall deck house, behind which there is much disorder and a large tarpaulin sheet, sloping from the roof to the hatch near the rounded stern.

On the foredeck stands a neat little dog kennel, complete with rough-haired terrier. Smaller canal boats all keep their dogs, usually terriers capable of squeezing their thin bodies into the smallest corner. The boat dog, looking rather like Dog Toby with the ruff of fur round his neck, is usually very proud of his boat and his responsibilities, and barks furiously at all comers. Before now I have seen one of them, enraged by the scornful air of a cat on the bank, bark with such vigour that he overbalanced and toppled into the canal.

The family in the *Ivy* are one of a number who live all the year around in their converted boats on the Foss Dyke and the Wytham. It may be necessity or it may be choice, but I suspect that they are not much more comfortable than those who dwelt in the barges which the Roman settlers built.

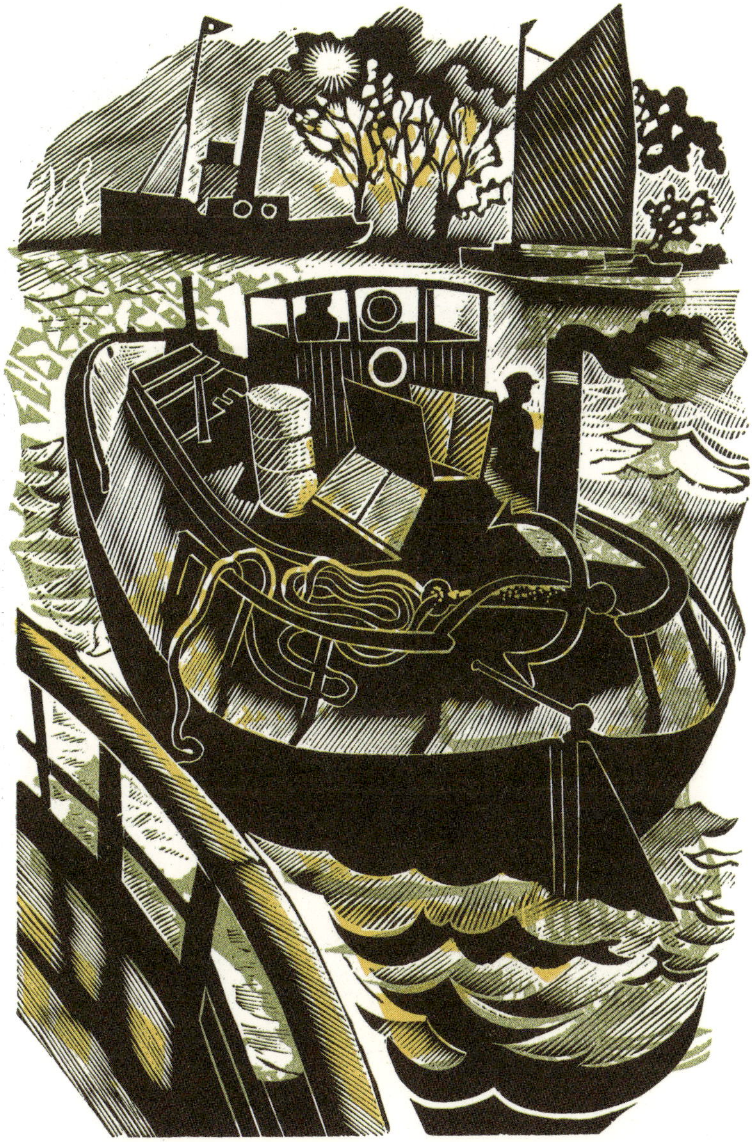

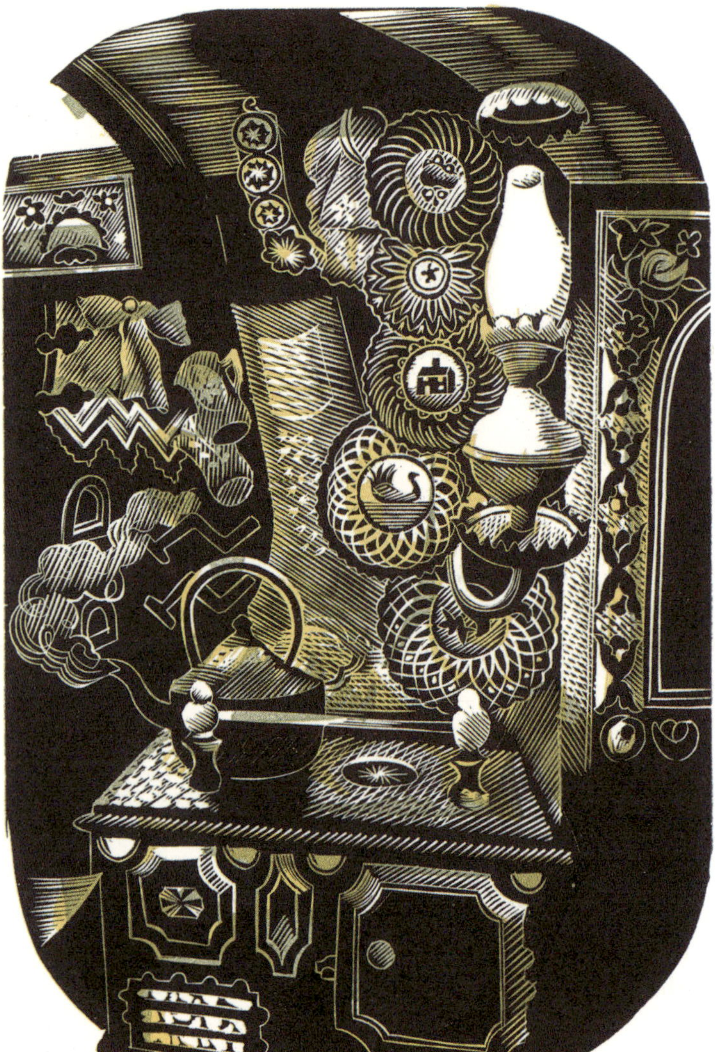

XII

ENDPIECE
NARROW BOAT IN DECEMBER

John and Lil were moving their household goods out of their boat into a temporary one at the yard wharf. John said, 'She's got nine holes in her now and one of them's below the waterline when she's loaded. If I can get this upper leak mended I can carry sugar.' He indicated a half-inch crack in the metal casing. 'I'm glad she's sprung it on us here,' he said. 'They'll do her up and paint her here better than anywhere else.' There was a compliment for the Uxbridge craftsmen!

'She's ready for it' said Lil. 'It's six years since she was painted, and it should be two, or three at the most.' She certainly needed it; the timbers of the cabin were cracked, and the damp had settled into the grain of the wood where the paint had peeled away.

Later, inside the cabin, 'That's where the draught comes up,' said Lil, holding her hand against the wall by the bunk. She

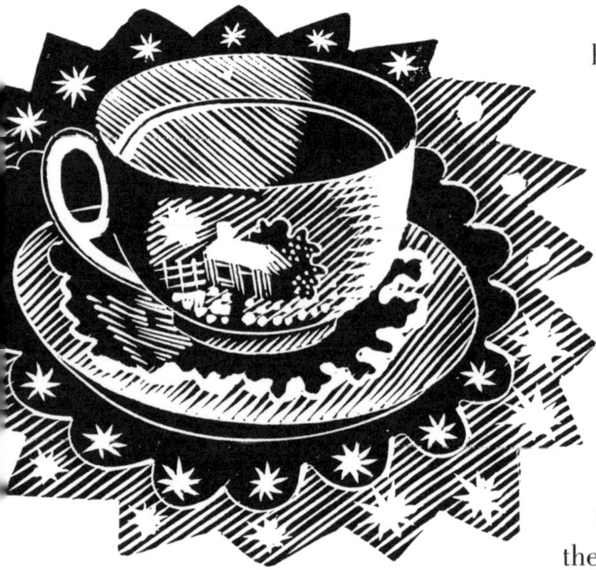

was busy packing up bedding for the move; the eleven-month baby slept soundly through all the disturbance.

It was nearly Christmas. 'We're going to Nottingham for Christmas,' said Lil. 'We've a long way to go, haven't we? I expect the boat'll be ready by the time we get back, and then we'll be busy.'

John slowly undid a roll of strong canvas, revealing an accurate painting, like a Chinese scroll, of the side view of a motor boat going into a lock. 'I done that,' he said, 'but I got stuck with the water.' I suggested a broken, wind-ruffled surface or a reflection of the boat. John looked out over the side. 'Ah, like it is now would do,' he said, 'just long lines. The boat, now. You don't often see that forward shape with a wooden frame to the awning. The new boats haven't got it. Oh well, when we get our boat back, I'll be able to paint some of the flower pieces myself.'

In the well of the boat something was rushing madly around and around. 'What sort of an animal do you call that?' I asked.

'You may well ask,' said John, 'he's a cross between a fox terrier and a whippet.' There was perhaps something of the whippet in his lean quarters and his back, but no particular breed of dog seemed to have gone to the making of the rest of him, though he had an appealing head and ears. He was barking furiously at a brown terrier on the bank, but the boat was just far enough out in the stream to keep the two apart.

There was a skin of ice on the wood and the dog slithered and skidded each time he turned. I asked John if he would get through to Nottingham in that weather. 'If it freezes up well,' he said, 'we'll have to get her into dock and go by train, unless we've a load on, of course – we have to stop with her then. I could tell you a story about freezing.'

He put his scroll in the drawer under the seat and patted the pile of blankets straight. Living in a tiny cabin within a few inches of the water has made him careful where he puts his possessions. Lil made up the stove while we talked. She was chasing the daylight, and it would be dark in an hour or two.

'My brother it was, in 1940,' said John. 'That's when we sunk the ice-boat. I was dredging then, before I married Lil. We were near Nottingham, and it began to freeze one Saturday. By Tuesday the ice was pretty thick and we got the ice-boat out, a thirty-foot iron one. We had eight horses on it and my brother steered. He wore waders up to here,' indicating his chin, 'and two jerseys and a waterproof jacket and thick boots to keep warm. We didn't know then that you had to stop towing and let her run up on the ice on her own. So up she came to the ice and rode up over the top for about a hundred yards. My brother was clinging to the tiller, but when we let her go she turned half over on her side, dropped into a crack in the ice and sank like a stone. We had to fish him out with a poling-hook. He stood shaking water and bits of ice out of his waders, and you should've heard his language. We laughed, but it wasn't so funny when we had to explain about the boat. ... Another time, Lil and I were near Rugby and it froze all round us overnight. I looked out in the morning and I could have walked off the boat on to the ice and we were stuck at an angle, too. It wasn't light, and the wood alongside was full of white trees in a fog . . . ghostly, they looked. It meant an eight-mile bike ride for me into Rugby for rations, until it snowed. Then I waded through it up to my waist across the fields to the high road. It was ten days or more till we could get away. Still, it was warm in the cabin at night. Not in the motor-cabin, though. It's asbestos-lined and tin-plated and it sweats.' I had heard before of a family who had a metal cabin and were unable to hang up any clothes at night because of the moisture which dripped from the walls. 'The wood keeps the heat in and the cold out,' said Lil, 'and it's cooler in summer too.'

By now, the bedding was packed and stored away, and the cups and saucers put in the cupboard. The baby was the last item in the move. The sun was low and red in the light fog and one's breath was clearly visible in the frosty air. At daybreak next morning they were off to Brentford – Lil called it Branford – to load up. I hoped there would be no severe frost for the next ten days, or a disappointed young family would spend their Christmas ice-bound in some remote and shallow Midland valley with neither cliff nor woods to break the force of a north-east wind.

I took my last look at the newly-fitted cabin, with the stove burning well, and the lamp lighted for the evening. As I turned away, the closing door shut off the bright lamp-light from the world, and I felt that these two people and their son were cut off from the world of today and had returned to some December long ago, when similar lamps were lighted beside an identical wharf.

IN ARCADIA

A series of beautifully illustrated books inspired by the many faces of the British countryside.

OTHER TITLES IN THIS SERIES

In the Heart of the Country • H.E. Bates

The genius of H.E. Bates and C.F. Tunncliffe combine to show the magnificence of winter blizzards; the sudden sweet arrival of spring blossom; the heavy scent of an August garden; or the bronze glow of autumn light.

The Happy Countryman • H.E. Bates

Inspired by the gentle beauty of the English landscape, the celebrated writer H.E. Bates and the much loved wildlife artist C.F. Tunnicliffe produced an enchanting tribute to country life.

Birds of the Hedgerow, Field and Woodland • Raphael Nelson

With a clever eye and subtle use of his engraving tools, Nelson manages to convey the minute differences between birds allowing the reader immediate recognition.

English Country Houses • Vita Sackville West

This book offers a brief history of the English country house from the Middle Ages to the 20th century, and of the people who built and lived in them – from common squires to kings and queens.

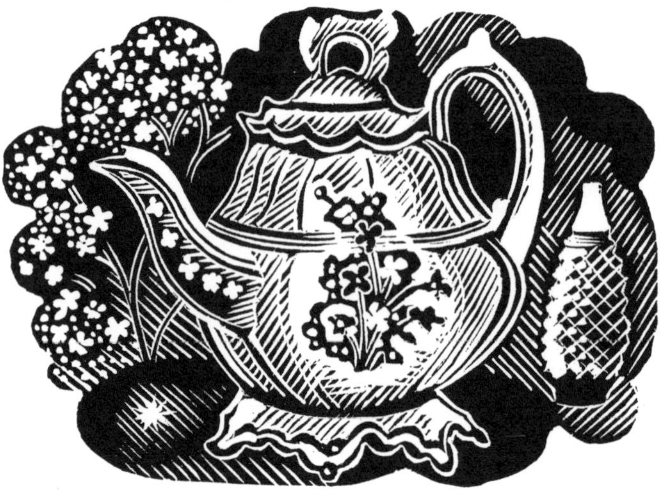